NanoScience and Technology

NanoScience and Technology

Series Editors:
P. Avouris B. Bhushan D. Bimberg K. von Klitzing H. Sakaki R. Wiesendanger

The series NanoScience and Technology is focused on the fascinating nano-world, mesoscopic physics, analysis with atomic resolution, nano and quantum-effect devices, nanomechanics and atomic-scale processes. All the basic aspects and technology-oriented developments in this emerging discipline are covered by comprehensive and timely books. The series constitutes a survey of the relevant special topics, which are presented by leading experts in the field. These books will appeal to researchers, engineers, and advanced students.

Please view available titles in *NanoScience and Technology* on series homepage
http://www.springer.com/series/3705/

Horst Hahn
Anatoli Sidorenko
Ion Tiginyanu

(Editors)

Nanoscale Phenomena

Fundamentals and Applications

With 106 Figures

Editors

Professor Dr. Horst Hahn
Forschungszentrum Karlsruhe GmbH
Institut für Nanotechnologie
76021 Karlsruhe, Germany
horst.hahn@int.fzk.de

Professor Dr. Anatoli Sidorenko
Professor Dr. Ion Tiginyanu
Academy of Science of Moldova
Institute of Electronic Engineering
and Industrial Technologies
Academiei Street 3/3, 2028 Chisinau, Moldova
sidorenko@int.fzk.de
tiginyanu@asm.md

Series Editors

Professor Dr. Phaedon Avouris
IBM Research Division
Nanometer Scale Science & Technology
Thomas J. Watson Research Center
P.O. Box 218
Yorktown Heights, NY 10598, USA

Professor Dr. Bharat Bhushan
Ohio State University
Nanotribology Laboratory
for Information Storage
and MEMS/NEMS (NLIM)
Suite 255, Ackerman Road 650
Columbus, Ohio 43210, USA

Professor Dr. Dieter Bimberg
TU Berlin, Fakutät Mathematik/
Naturwissenschaften
Institut für Festkörperphyisk
Hardenbergstr. 36
10623 Berlin, Germany

Professor Dr., Dres. h.c. Klaus von Klitzing
Max-Planck-Institut
für Festkörperforschung
Heisenbergstr. 1
70569 Stuttgart, Germany

Professor Hiroyuki Sakaki
University of Tokyo
Institute of Industrial Science
4-6-1 Komaba, Meguro-ku
Tokyo 153-8505, Japan

Professor Dr. Roland Wiesendanger
Institut für Angewandte Physik
Universität Hamburg
Jungiusstr. 11
20355 Hamburg, Germany

NanoScience and Technology ISSN 1434-4904
ISBN 978-3-642-00707-1   e-ISBN 978-3-642-00708-8
DOI 10.1007/978-3-642-00708-8
Springer Heidelberg Dordrecht London New York

Library of Congress Control Number: 2009929172

© Springer-Verlag Berlin Heidelberg 2009

This work is subject to copyright. All rights are reserved, whether the whole or part of the material is concerned, specifically the rights of translation, reprinting, reuse of illustrations, recitation, broadcasting, reproduction on microfilm or in any other way, and storage in data banks. Duplication of this publication or parts thereof is permitted only under the provisions of the German Copyright Law of September 9, 1965, in its current version, and permission for use must always be obtained from Springer. Violations are liable to prosecution under the German Copyright Law.

The use of general descriptive names, registered names, trademarks, etc. in this publication does not imply, even in the absence of a specific statement, that such names are exempt from the relevant protective laws and regulations and therefore free for general use.

Cover design: SPi Publisher Services

Printed on acid-free paper

Springer is part of Springer Science+Business Media (www.springer.com)

Preface

Nanotechnology – a rapidly developing modern area of science & technology, covered many areas of physics, chemistry, materials science, and biology over the last decade. The unique properties of materials at the nanometer scale and the outstanding performance of the nanoscale devices are the main reasons for the immense growth of this field. Nanotechnological findings serve as the base for enormous developments of electronics and many new branches, such as spintronics and single-electron devices, new approaches for medical treatment and diagnostic procedures, the implementation of high-tech sensors and actuators. Moreover, nanotechnology does not only offer novel products but also initiates new areas, such as photonics and metamaterials. Of course, in one book, it is practically impossible to present a comprehensive overview of all areas of nanotechnology. The main goal of the present book is to show an intrinsic correlation and mutual influence of three important parts of nanoscience: new phenomena – nanomaterials – nanodevices. For the discovery of new phenomena, it is necessary to develop novel nanotechnological processes for the fabrication of nanomaterials. The nanostructures and new phenomena serve as the base for the development of novel nanoelectronic devices.

According to this concept, we organized the book into 5 parts –
Coherent Effects in Nanostructures
Nanomaterials and Nanoparticles
Nanoelectronics
Nanobiology
Philosophical Aspects of Nanoscience
presenting thoroughly selected articles reported at the International Symposium *"Nanoscale Phenomena – Fundamentals and Applications"* (Chisinau, September 19-22, 2007). The symposium brought together leading experts – experimentalists, theorists, and engineers – working in nanoscience and nanotechnology with the aim to share their expertise and experience on how the new fundamental ideas and principles can be rapidly implemented in the areas mentioned above.

The symposium was financially supported by the Alexander von Humboldt Foundation as a "Humboldt Kolleg", and the editors are grateful for this generous support.

June 2009

Horst Hahn
Anatoli Sidorenko
Ion Tiginyanu

Contents

Part I Coherent Effects in Nanostructures

1 Extinction and Recovery of Superconductivity by Interference in Superconductor/Ferromagnet Bilayers 3
A.S. Sidorenko, V.I. Zdravkov, J. Kehrle, R. Morari,
E. Antropov, G. Obermeier, S. Gsell, M. Schreck, C. Müller,
V.V. Ryazanov, S. Horn, R. Tidecks, and L.R. Tagirov
  1.1 Introduction 3
  1.2 Sample Preparation and Characterization 6
  1.3 Results of Superconducting T_C Measurements and Discussion 8
  1.4 Conclusions 9
  References 10

2 Aharonov–Bohm Oscillations in Small Diameter Bi Nanowires 13
L. Konopko
  2.1 Introduction 13
  2.2 Experimental 15
  2.3 Results and Discussion 16
  2.4 Conclusions 19
  References 19

3 Point-Contact Study of the Superconducting Gap in the Magnetic Rare-Earth Nickel-Borocarbide $RNi_2B_2C(R = Dy, Ho, Er, Tm)$ Compounds 21
Yu.G. Naidyuk, N.L. Bobrov, V.N. Chernobay,
S.-L. Drechsler, G. Fuchs, O.E. Kvitnitskaya, D.G. Naugle,
K.D.D. Rathnayaka, L.V. Tyutrina, and I.K. Yanson
  3.1 Introduction 21
  3.2 Experimental 22
  3.3 Results and Discussion 22
  3.4 Conclusions 26
  References 27

4 Peculiarities of Supershort Light Pulses Transmission by Thin Semiconductor Film in Exciton Range of Spectrum 29
P.I. Khadzhi, I.V. Beloussov, D.A. Markov, A.V. Corovai, and V.V. Vasiliev
- 4.1 Introduction 29
- 4.2 Basic Equations 30
- 4.3 Discussion of Results of Numerical Solutions 33
- 4.4 Conclusions 37
- References 37

Part II Nanomaterials and Nanoparticles

5 Nanostructuring and Dissolution of Cementite in Pearlitic Steels During Severe Plastic Deformation 41
Y. Ivanisenko, X. Sauvage, I. MacLaren, and H.-J. Fecht
- 5.1 Introduction 41
- 5.2 Experimental 42
- 5.3 Results and Discussion 44
  - 5.3.1 Changes in the Microstructure and in Phase Composition of the Pearlitic Steel During HPT 44
  - 5.3.2 Variations of the Chemical Composition of Carbides 47
  - 5.3.3 Distribution of Released Carbon Atoms in the Microstructure 51
  - 5.3.4 Role of the Cementite Morphology 52
  - 5.3.5 Driving Force and Mechanism of Strain Induced Decomposition of Cementite 52
- 5.4 Conclusions 54
- References 54

6 Advanced Method for Gas-Cleaning from Submicron and Nanosize Aerosol 57
A. Bologa, H.-R. Paur, and H. Seifert
- 6.1 Introduction 57
- 6.2 Development of the Method and Electrostatic Precipitator 58
- 6.3 Influence of Gas Temperature on Current–Voltage Characteristics 60
- 6.4 Precipitation of Al_2O_3 Particles 60
- 6.5 Precipitation of TIO_2 Particles 63
- 6.6 Conclusion 64
- References 65

7 Deformation Microstructures Near Vickers Indentations in SNO_2/SI Coated Systems 67
G. Daria, H. Evghenii, S. Olga, D. Zinaida, M. Iana, and Z. Victor
- 7.1 Introduction 67

	7.2	Experimental	68
	7.3	Results and Discussion	68
	7.4	Conclusions	73
	References		74

8 Grain Boundary Phase Transformations in Nanostructured Conducting Oxides 75
B.B. Straumal, A.A. Myatiev, P.B. Straumal, and A.A. Mazilkin

	8.1	Introduction	75
	8.2	Grain Boundary Phase Transformations and Phase Diagrams	76
	8.3	Grain Boundary Phases in Zinc Oxide	77
	8.4	Conducting Oxides of Fluorite Structure	80
		8.4.1 GB Wetting Phases	80
		8.4.2 Monolayer GB Segregation	81
		8.4.3 Scavengers for GB Impurities	82
		8.4.4 Heavy Doping	83
	8.5	GB Phenomena in Perovskites	84
	8.6	Influence of Synthesis Route on the Properties of Nanostructured Materials	84
	8.7	Synthesis of Nanostructured Oxides by a "Liquid Ceramics" Method	85
	8.8	Conclusions	86
	References		87

9 Copper Electrodeposition from Ultrathin Layer of Electrolyte 89
S. Zhong, T. Koch, M. Wang, M. Zhang, and T. Schimmel

	9.1	Introduction	89
	9.2	Experimental Methods	90
		9.2.1 Copper Submicrowires	92
		9.2.2 Periodically Nanostructured Films	97
	9.3	Conclusion	100
	References		100

10 Effect of Plasma Environment on Synthesis of Vertically Aligned Carbon Nanofibers in Plasma-Enhanced Chemical Vapor Deposition 103
Igor Denysenko, Kostya Ostrikov, Nikolay A. Azarenkov, and Ming Y. Yu

	10.1	Introduction	103
	10.2	Theoretical Model	104
	10.3	Results and Discussion	107
	10.4	Conclusions	109
	References		110

Part III Nanoelectronics

11 Single-Atom Transistors: Switching an Electrical Current with Individual Atoms 113
Christian Obermair, Fangqing Xie, Robert Maul, Wolfgang Wenzel, Gerd Schön, and Thomas Schimmel
- 11.1 Introduction 113
- 11.2 Experimental 114
- 11.3 Configuring a Bistable Atomic Switch by Repeated Electrochemical Cycling 116
- 11.4 Preselectable Integer Quantum Conductance of Electrochemically Fabricated Silver Point Contacts 118
- 11.5 Summary 121
- References 122

12 Electronically Tunable Nanostructures: Metals and Conducting Oxides 125
Subho Dasgupta, Robert Kruk, and Horst Hahn
- 12.1 Introduction 125
- 12.2 Tunable Change in Electronic Transport of a Metal 130
  - 12.2.1 Nanoporous Gold Electrode from De-alloying 130
  - 12.2.2 Variation in Resistance in Thin Gold Film Electrode 130
- 12.3 Reversible Change in Electronic Transport in a High Conducting Transparent Oxide Nanoparticulate Thin Film 133
- 12.4 Summary 136
- References 137

13 Impedance Spectroscopy as a Powerful Tool for Better Understanding and Controlling the Pore Growth Mechanism in Semiconductors 139
A. Cojocaru, E. Foca, J. Carstensen, M. Leisner, I.M. Tiginyanu, and H. Föll
- 13.1 Introduction 139
- 13.2 Experimental 140
- 13.3 Results and Discussion 140
- 13.4 Conclusions 143
- References 144

14 Studying Functional Electrode Structures with Combined Scanning Probe Techniques 145
P. Dupeyrat, M. Müller, R. Gröger, Th. Koch, C. Eßmann, M. Barczewski, and Th. Schimmel
- 14.1 Introduction 145
- 14.2 AFM Characterization and Grain Size Analysis 147
- 14.3 Chemical Contrast Imaging 147
- 14.4 Electrostatic Force Microscopy (EFM) 150

14.5	Implementation and Test of the EFM Method	152
14.6	Electrical Characterization of 8YSZ-MOD Layers	154
14.7	Summary	157
References		158

Part IV Nanobiology

15 Integrated Lab-on-a-Chip System in Life Sciences 161
S. Thalhammer, M.F. Schneider, and A. Wixforth

15.1	Lab-on-a-Chip Systems	161
15.2	General Manipulation of Cells and Cell Components in Microdevices	163
15.3	Actuation of Lab-on-a-Chip Systems	165
15.4	Lab-on-a-Chip Concepts	168
15.5	Acoustically Driven Microfluidics	169
15.6	Experimental Details	170
15.7	Acoustic Mixing	171
15.8	Droplet Actuation	172
15.9	PCR-Chips	172
15.10	Stationary On-Chip PCR	173
15.11	PCR on a Chip	176
15.12	Blood Flow on a Chip	177
15.13	Proteins Under Flow	178
15.14	Cell–Cell Interactions on a Chip	180
15.15	Microdissection	180
15.16	Extended Glass-Needle Microdissection	181
15.17	Laser-Based Microdissection	182
15.18	Atomic Force Microscopy Microdissection	183
15.19	Acoustically Driven Cytogenetic Lab-on-a-Chip	184
15.20	Summary	186
References		187

Part V Philosophical Aspects of Nanoscience

16 Methodological Problems of Nanotechnoscience 193
V.G. Gorokhov

16.1	Different Definitions of Nanotechnology	194
16.2	Nanotheory as a Cluster of the Different Natural and Engineering Theories	195
16.3	Nano Systems Engineering	201

Index 207

Contributors

E. Antropov Institute of Electronic Engineering and Industrial Technologies, MD-2028 Kishinev, Moldova

N.A. Azarenkov School of Physics and Technology, V. N. Karazin Kharkiv National University, 4 Svobody sq., 61077 Kharkiv, Ukraine

M. Barczewski Institute of Nanotechnology, Forschungszentrum Karlsruhe GmbH, P.O. Box 3640, D-76021 Karlsruhe, Germany

I.V. Beloussov Institute of Applied Physics, Academy of Science of Moldova, MD-2028 Chisinau, Republic of Moldova

N.L. Bobrov B. Verkin Institute for Low Temperature Physics and Engineering (ILTPE), National Academy of Sciences of Ukraine, 47 Lenin avenue, 61103, Kharkiv, Ukraine
and
Department of Physics, Texas A&M University, College Station, TX 77843-4242, USA

A. Bologa Forschungszentrum Karlsruhe, Institute fürTechnische Chemie, 76021, Karlsruhe, Germany

J. Carstensen Institute for Materials Science, Christian-Albrechts-University of Kiel, D-24143 Kiel, Germany

V.N. Chernobay B. Verkin Institute for Low Temperature Physics and Engineering (ILTPE), National Academy of Sciences of Ukraine, 47 Lenin avenue, 61103, Kharkiv, Ukraine

A. Cojocaru Institute for Materials Science, Christian-Albrechts-University of Kiel, D-24143 Kiel, Germany

A.V. Corovai Dniester State University, Tiraspol, MD 3300, Moldova

Z. Danitsa Institute of Applied Physics, Academy of Sciences of Moldova, 5 Academy street, MD-2028, Chisinau, Moldova

S. Dasgupta Institute of Nanotechnology, Forschungszentrum Karlsruhe GmbH, P.O. Box 3640, D-76021 Karlsruhe, Germany

I. Denysenko School of Physics and Technology, V. N. Karazin Kharkiv National University, 4 Svobody sq., 61077 Kharkiv, Ukraine
and
Plasma Nanoscience, Complex Systems, School of Physics, The University of Sydney, Sydney, New South Wales 2006, Australia; CSIRO Materials Science and Engineering, Lindfield NSW 2070, Australia

S.-L. Drechsler Leibniz-Institut für Festkörper- und Werkstoffforschung (IFW) Dresden, POB 270116, D-01171 Dresden, Germany

P. Dupeyrat Institute of Nanotechnology, Forschungszentrum Karlsruhe GmbH, P.O. Box 3640, D-76021 Karlsruhe, Germany

C. Eßmann Institute of Applied Physics, Universität Karlsruhe, D-76128 Karlsruhe, Germany

H.-J. Fecht Universität Ulm, Institute of Micro and Nanomaterials, Albert-Einstein-Allee-47, 89081 Ulm, Germany

E. Foca Institute for Materials Science, Christian-Albrechts-University of Kiel, D-24143 Kiel, Germany

H. Föll Institute for Materials Science, Christian-Albrechts-University of Kiel, D-24143 Kiel, Germany

G. Fuchs Leibniz-Institut für Festkörper- und Werkstoffforschung (IFW) Dresden, POB 270116, D-01171 Dresden, Germany

V.G. Gorokhov Institute for Philosophy, Russian Academy of Sciences, Moscow, Russia
and
Institute of Technology Assessment and Systems Analysis, Forschungszentrum Karlsruhe, D-76021 Karlsruhe, Germany

D. Grabco Institute of Applied Physics, Academy of Sciences of Moldova, 5 Academy street, MD-2028, Chisinau, Moldova

R. Gröger Institute of Nanotechnology, Forschungszentrum Karlsruhe GmbH, P.O. Box 3640, D-76021 Karlsruhe, Germany

S. Gsell Institut für Physik, Universität Augsburg, D-86159 Augsburg, Germany

H. Hahn Institute of Nanotechnology, Forschungszentrum Karlsruhe GmbH, P.O. Box 3640, D-76021 Karlsruhe, Germany
and
Joint Research Laboratory Nanomaterials, Technische Universität Darmstadt, Institute of Materials Science, Petersenstr. 23, 64287 Darmstadt, Germany

E. Harea Institute of Applied Physics, Academy of Sciences of Moldova, 5 Academy street, MD-2028, Chisinau, Moldova

S. Horn Institut für Physik, Universität Augsburg, D-86159 Augsburg, Germany

Contributors

Y. Ivanisenko Institute of Nanotechnology, Forschungszentrum Karlsruhe, D-76021 Karlsruhe, Germany

J. Kehrle Institut für Physik, Universität Augsburg, D-86159 Augsburg, Germany

P.I. Khadzhi Institute of Applied Physics, Academy of Science of Moldova, MD-2028 Chisinau, Republic of Moldova

Th. Koch Chair of Particle Technology, Friedrich-Alexander-University Erlangen-Nuremberg, D-91058 Erlangen, Germany
and
Institute of Nanotechnology, Forschungszentrum Karlsruhe GmbH, P.O. Box 3640, D-76021 Karlsruhe, Germany

L. Konopko Institute of Electronic Engineering and Industrial Technologies, ASM, Chisinau, Moldova
and
International Laboratory of High Magnetic Fields and Low Temperatures, Wroclaw, Poland

R. Kruk Institute of Nanotechnology, Forschungszentrum Karlsruhe GmbH, P.O. Box 3640, D-76021 Karlsruhe, Germany

O.E. Kvitnitskaya B. Verkin Institute for Low Temperature Physics and Engineering (ILTPE), National Academy of Sciences of Ukraine, 47 Lenin avenue, 61103, Kharkiv, Ukraine
and
Leibniz-Institut für Festkörper- und Werkstoffforschung (IFW) Dresden, P.O. Box 270116, D-01171 Dresden, Germany

M. Leisner Institute for Materials Science, Christian-Albrechts-University of Kiel, D-24143 Kiel, Germany

I. MacLaren Department of Physics and Astronomy, University of Glasgow, Glasgow G12 8QQ, UK

D.A. Markov Dniester State University, Tiraspol, MD 3300, Moldova

R. Maul Institut für Nanotechnologie, Forschungszentrum Karlsruhe GmbH, 76021 Karlsruhe, Germany

A.A. Mazilkin Institute of Solid State Physics RAS, 142432 Chernogolovka, Russia

I. Mirgorodscaia Institute of Applied Physics, Academy of Sciences of Moldova, 5 Academy street, MD-2028, Chisinau, Moldova

R. Morari Institute of Electronic Engineering and Industrial Technologies, MD2028 Kishinev, Moldova

C. Müller Institut für Physik, Universität Augsburg, D-86159 Augsburg, Germany

M. Müller Institute of Applied Physics, Universität Karlsruhe, D-76128 Karlsruhe, Germany

A.A. Myatiev Moscow Institute of Steel and Alloys, 119049 Moscow, Russia

Yu.G. Naidyuk B. Verkin Institute for Low Temperature Physics and Engineering (ILTPE), National Academy of Sciences of Ukraine, 47 Lenin avenue, 61103, Kharkiv, Ukraine
and
Leibniz-Institut für Festkörper- und Werkstoffforschung (IFW) Dresden, POB 270116, D-01171 Dresden, Germany

D.G. Naugle Department of Physics, Texas A&M University, College Station, TX 77843-4242, USA

Ch. Obermair Institut für Angewandte Physik, Universität Karlsruhe, 76128 Karlsruhe, Germany
and
DFG-Center for Functional Nanostructure (CFN), Universität Karlsruhe, 76128 Karlsruhe, Germany

G. Obermeier Institut für Physik, Universität Augsburg, D-86159 Augsburg, Germany

K. Ostrikov Plasma Nanoscience, Complex Systems, School of Physics, The University of Sydney, Sydney, New South Wales 2006, Australia; CSIRO Materials Science and Engineering, Lindfield NSW 2070, Australia

H.-R. Paur Forschungszentrum Karlsruhe, Institute fürTechnische Chemie, 76021, Karlsruhe, Germany

K.D.D. Rathnayaka Department of Physics, Texas A&M University, College Station, TX 77843-4242, USA

V.V. Ryazanov Institute of Solid State Physics, Russian Academy of Sciences, 142432 Chernogolovka, Russia

X. Sauvage University of Rouen, Groupe de Physique des Matériaux, CNRS (UMR 6634), F-76801 Saint-Etienne du Rouvray, France

Th. Schimmel Institut für Angewandte Physik, Universität Karlsruhe, 76128 Karlsruhe, Germany
and
DFG-Center for Functional Nanostructure (CFN), Universität Karlsruhe, 76128 Karlsruhe, Germany
and
Institut für Nanotechnologie, Forschungszentrum Karlsruhe GmbH, 76021 Karlsruhe, Germany

M.F. Schneider University of Augsburg, Experimental Physics I, D-86159 Augsburg, Germany

Contributors

G. Schön Institut für Theoretische Festkörperphysik, Universität Karlsruhe, 76128 Karlsruhe, Germany
and
DFG-Center for Functional Nanostructure (CFN), Universität Karlsruhe, 76128 Karlsruhe, Germany
and
Institut für Nanotechnologie, Forschungszentrum Karlsruhe GmbH, 76021 Karlsruhe, Germany

M. Schreck Institut für Physik, Universität Augsburg, D-86159 Augsburg, Germany

H. Seifert Forschungszentrum Karlsruhe, Institute fürTechnische Chemie, 76021, Karlsruhe, Germany

O. Shikimaka Institute of Applied Physics, Academy of Sciences of Moldova, 5 Academy street, MD-2028, Chisinau, Moldova

A.S. Sidorenko Institute of Electronic Engineering and Industrial Technologies, MD2028 Kishinev, Moldova

B.B. Straumal Institute of Solid State Physics RAS, 142432 Chernogolovka, Russia
and
Moscow Institute of Steel and Alloys, 119049 Moscow, Russia

P.B. Straumal Moscow Institute of Steel and Alloys, 119049 Moscow, Russia

L.R. Tagirov Solid State Physics Department, Kazan State University, 420008 Kazan, Russia

S. Thalhammer Helmholtz Zentrum Muenchen, German Research Center for Environmental Health, D-85764 Neuherberg, Germany

R. Tidecks Institut für Physik, Universität Augsburg, D-86159 Augsburg, Germany

I.M. Tiginyanu Institute of Applied Physics, Academy of Science of Moldova, MD-2028 Chisinau, Republic of Moldova

L.V. Tyutrina B. Verkin Institute for Low Temperature Physics and Engineering (ILTPE), National Academy of Sciences of Ukraine, 47 Lenin avenue, 61103, Kharkiv, Ukraine

V.V. Vasiliev Dniester State University, Tiraspol, MD 3300, Moldova

M. Wang National Laboratory of Solid State Microstructures and Department of Physics, Nanjing University, Nanjing 210093, China

W. Wenzel Institut für Nanotechnologie, Forschungszentrum Karlsruhe GmbH, 76021 Karlsruhe, Germany
and
DFG-Center for Functional Nanostructure (CFN), Universität Karlsruhe, 76128 Karlsruhe, Germany

A. Wixforth University of Augsburg, Experimental Physics I, D-86159 Augsburg, Germany

F.Q. Xie Institut für Angewandte Physik, Universität Karlsruhe, 76128 Karlsruhe, Germany
and
DFG-Center for Functional Nanostructure (CFN), Universität Karlsruhe, 76128 Karlsruhe, Germany

I.K. Yanson B. Verkin Institute for Low Temperature Physics and Engineering (ILTPE), National Academy of Sciences of Ukraine, 47 Lenin avenue, 61103, Kharkiv, Ukraine

M.Y. Yu Institute for Fusion Theory and Simulation, Department of Physics, Zhejiang University, 310027 Hangzhou, China
and
Theoretical Physics I, Ruhr University, D-44780 Bochum, Germany

V. Zalamai Institute of Applied Physics, Academy of Sciences of Moldova, 5 Academy street, MD-2028, Chisinau, Moldova

V.I. Zdravkov Institute of Electronic Engineering and Industrial Technologies, MD2028 Kishinev, Moldova

M. Zhang National Laboratory of Superhard Materials and Institute of Atomic and Molecular Physics, Jilin University, Changchun 130012, China

Sh. Zhong Institute of Nanotechnology, Forschungszentrum Karlsruhe, D-76021 Karlsruhe, Germany

Part I
Coherent Effects in Nanostructures

Chapter 1
Extinction and Recovery of Superconductivity by Interference in Superconductor/Ferromagnet Bilayers

A.S. Sidorenko, V.I. Zdravkov, J. Kehrle, R. Morari, E. Antropov,
G. Obermeier, S. Gsell, M. Schreck, C. Müller, V.V. Ryazanov, S. Horn,
R. Tidecks, and L.R. Tagirov

Abstract In superconductor–ferromagnet (S/F) metallic contacts, the superconducting condensate penetrates through the S/F interface into a ferromagnetic layer. In contrast to the conventional S/N proximity effect, the pairing wave function not only decays deep into the F metal, but simultaneously oscillates. Interference of the oscillating pairing function in a ferromagnetic film gives rise to a modulation of the pairing function flux crossing the S/F interface, which results in oscillations of superconducting transition temperature of the adjacent S layer. In this work, we report on the experimental observation of the superconductivity reentrance phenomenon with double suppression of the superconductivity in Nb/Cu$_{1-x}$Ni$_x$ bilayers as a function of the ferromagnetic layer thickness, d_{CuNi}. The superconducting T_c drops sharply with increasing d_{CuNi} till total suppression of superconductivity at $d_{\text{CuNi}} \approx 2.5$ nm. At a further increase of the Nb/Cu$_{1-x}$Ni$_x$ layer thickness, the superconductivity restores at $d_{\text{CuNi}} \geq 24$ nm. Then, with the subsequent increase of d_{CuNi}, the superconductivity vanishes again at $d_{\text{CuNi}} \approx 38$ nm.

1.1 Introduction

In superconductor–ferromagnetic metal (S/F) contacts, the superconducting pairing wave function not only exponentially decays into the F metal, as in the superconductor/normal metal (S/N) proximity effect [1, 2], but simultaneously oscillates [3, 4]. A variety of novel physical effects caused by these oscillations was predicted (see reviews [5–8] and references therein). Some of them have already been observed experimentally: nonmonotonous behavior of the superconducting critical temperature, T_c, as a function of the F metal layer thickness [9–13], Josephson junctions with intrinsic π-phase shift across the junction [14], and inverted, cap-sized differential current–voltage characteristics [15]. In this work, we report on results of observation of the reentrant T_c phenomenon with double suppression of superconductivity in Nb/Cu$_{1-x}$Ni$_x$ bilayers ($x = 0.59$) for increasing ferromagnetic Cu$_{1-x}$Ni$_x$ layer thickness, d_{CuNi}. After a destruction by interference effects of the superconducting pairing wave function and a subsequent recovery, a second suppression of

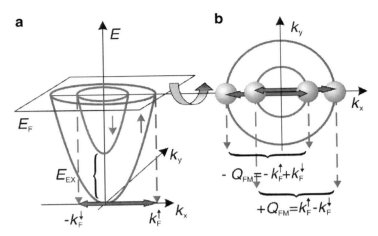

Fig. 1.1 Origin of the FFLO state. (**a**) Spin-splitting E_{ex} of the conduction band of a ferromagnet by the exchange field. Sketch for $k_z = 0$. (**b**) Cross-section of the band energy dispersions for $k_z = 0$ at the Fermi energy. Paired electrons (*green with red balls*) establish from the majority (*green*) and minority (*red*) subbands (wave number vectors indicated in the respective color). The FFLO pairing momentum along the x axis is $\hbar Q_{FM} = \hbar \Delta k_F = E_{ex}/v_F$

superconductivity is found, giving an impressive experimental evidence for a quasi-one dimensional Fulde–Ferrell–Larkin–Ovchinnikov (FFLO) [16, 17] like state in the ferromagnetic layer.

At a plane S/F interface, the quasi-one-dimensional FFLO-like state can be generated in the F material [3–8]. Due to the exchange splitting of the conduction band (Fig. 1.1a), one of the singlet Cooper-pair electrons occupies the majority subband, e.g., spin-up, while the other one resides at the spin-down, minority subband (Fig. 1.1b). Although the pairing occurs with opposite directions of the wave number vectors of the electrons, their absolute values are not equal due to the exchange splitting of the conduction band (see Fig. 1.1a). The resulting pairing state acquires a finite momentum of $\hbar Q_{FM} = E_{ex}/v_F$, where $E_{ex} \ll E_F$ is the energy of the exchange splitting of a free-electron-like, parabolic conduction band, E_F is the Fermi energy, and v_F is the Fermi velocity. Then, the pairing function of this state does not simply decay as it would be in a nonmagnetic metal, but oscillates on a wavelength scale λ_{FM} (i.e., $\lambda_{FM} = 2\pi/k_{FM}$) given by the magnetic coherence length ξ_F. In a clean ferromagnet ($l_F \gg \xi_{F0}$), it is $\lambda_{F0} \equiv 2\pi \xi_{F0} = 2\pi \hbar v_F/E_{ex}$ [4, 18], whereas in the dirty case ($l_F \ll \xi_{F0}$), we get $\lambda_{FD} = 2\pi \xi_{FD} = 2\pi (2\hbar D_F/E_{ex})^{1/2}$ [3, 7], where $D_F = l_F v_F/3$ with l_F the electron mean free path in the F-metal. The decay length of the pairing wave function is l_F and ξ_{FD} in the clean and dirty cases respectively [3, 4, 19].

The oscillation of the pairing wave function in the F-metal is the reason for an oscillatory S/F proximity effect, yielding a nonmonotonous, oscillating dependence of the superconducting critical temperature, T_c, on the ferromagnetic layer thickness, d_F. The phenomenon can be qualitatively described using the analogy

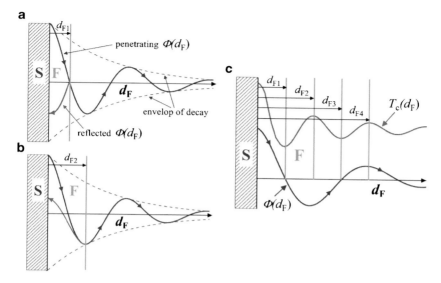

Fig. 1.2 Interference of the pairing function Φ in an S/F bilayer at the S/F interface assuming the case that $1/2$ of the pairing function amplitude is reflected at the boundary, and $1/2$ penetrates into the F material. Spin-dependent phase-shifts of the penetrating pairing function at the S/F interface [21–23] as well as phase shifts for the reflected waves are neglected for simplicity. (**a**) For $d_{F1} \sim \lambda_{FM}/4$ the resulting amplitude is minimal. (**b**) For $d_{F2} \sim \lambda_{FM}/2$ the subsequent local maximum of the amplitude if reached for increasing d_F. (**c**) Modulation of the superconducting transition temperature T_c of a thin superconducting film correlated with the interference conditions

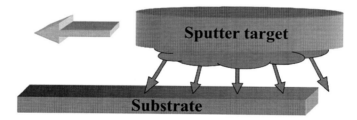

Fig. 1.3 The moving target setup utilizing the spray deposition technique

with the interference of light in a parallel-sided plate of glass with a mirror coated back side, at normal incidence [20]. As the interference conditions change periodically between constructive and destructive upon changing the thickness of the plate, the flux of light through the interface of incidence is modulated. In a layered S/F system, the pairing function flux crossing the S/F interface depends on the ferromagnetic layer thickness, d_F, because of the pairing function interference (see Fig. 1.2a, b). As a result, the coupling between the S and F layers in a series of samples with increasing d_F is modulated, and the superconducting T_c oscillates as a function of d_F (Fig. 1.2c). The amplitude of T_c oscillation depends sensitively on the superconducting layer thickness (see discussion of Fig. 1.3a, b).

Recently, expressed oscillations and pronounced reentrance, i.e., an extinction and recovery of superconductivity as a function of d_F, were measured in a Nb/Cu$_{1-x}$Ni$_x$ bilayer [13]. However, the most spectacular evidence for the oscillatory proximity effect would be the observation of the multiple reentrant behavior of the superconducting state predicted theoretically [19, 24, 25]. To realize this regime experimentally, one has to study at first the $T_c(d_S)$ dependence for a series of S/F bilayers with constant d_F to find the range of the superconducting layer thickness, d_S, in which superconductivity is most sensitive to the destructive influence of the ferromagnetism.

1.2 Sample Preparation and Characterization

To fabricate the S/F bilayers, we used niobium as superconducting material and Cu$_{1-x}$Ni$_x$ ($x \approx 0.59$) alloy as ferromagnetic layer. The choice of the alloy instead of a conventional elemental ferromagnet has the following advantages for the experimentalist. The oscillation length $\lambda_{F0} = 2\pi \hbar v_F / E_{ex}$ in strong clean ferromagnets, like iron, nickel, or cobalt, is extremely short, because the exchange splitting energy, E_{ex}, is usually in the range 0.1–1.0 eV [9, 10, 12, 26]. Thus, to detect an oscillatory behavior of T_c experimentally, d_F must be very small, e.g., between 0.6 and 2.5 nm for pure Ni [26]. Weak ferromagnets with an order of magnitude smaller exchange splitting of the conduction band allow the observation of the effect at much larger thicknesses d_F of about 2–25 nm, which can be controlled and characterized more easily. Moreover, for a long-wavelength oscillation, the atomic-scale interface roughness does not any longer have a dominating effect on an extinction of T_c oscillations.

The S/F samples were prepared by magnetron sputtering on commercial (111) silicon substrates at room temperature. Three targets, Si, Nb, and Cu$_{40}$Ni$_{60}$ (75 mm in diameter), were presputtered for 10–15 min to remove contaminations and reduce the residual gas pressure (by Nb as getter material) in the chamber. First, a silicon buffer layer was deposited using a RF magnetron to generate a clean interface for the subsequently deposited Cu$_{1-x}$Ni$_x$ or niobium layer.

To prepare samples with variable thickness of one of the layers, a wedge-shaped film was deposited [13, 26]: the 80 mm long and 7 mm wide silicon substrate was mounted at a distance of 4.5 cm from the target symmetry axis to utilize the intrinsic spatial gradient of the deposition rate. The Cu$_{40}$Ni$_{60}$ target was RF sputtered with a rate 3–4 nm sec^{-1}.

To obtain flat, high-quality Nb layers with thicknesses in the range of 5–15 nm, we moved the full-power operating magnetron along the silicone substrate using the motorized setup (see Fig. 1.3). Thus, the surface was uniformly sprayed with the material, and the average deposition rate of the Nb film could be decreased down to 1.3 nm sec^{-1}, while for a fixed, nonmoving target, it would be about 4 nm sec^{-1}. To prevent degradation in an ambient atmosphere, the resulting Cu$_{1-x}$Ni$_x$/Nb or Nb/Cu$_{1-x}$Ni$_x$ bilayers were coated by an amorphous silicon cap of about 10 nm

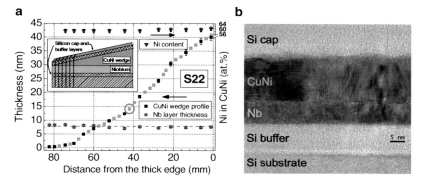

Fig. 1.4 Sample design and characterization. (**a**) RBS results for the thickness of the Nb and CuNi layers and Ni content in the CuNi alloy. Sketch of the layers stack, see the inset. Black rectangular symbols for CuNi alloy layer are measured points, orange symbols were linearly interpolated. (**b**) Transmission electron microscopy (TEM) cross-sectional image of a cut across the layers (sample S22–18 marked by a red circle in the left panel: $d_{Nb} \approx 7.8$ nm, $d_{CuNi} \approx 14.$ nm according to RBS)

thickness, which is insulating at low temperatures. After cutting the wedge samples into strips across the thickness gradient (see the inset in Fig. 1.4a), aluminum wires 50 μm in diameter were attached to the strips by an ultrasonic bonder for four-probe resistance measurements. For further sample preparation details, see [13, 26].

In the first kind of samples, the superconducting Nb layer was of variable thickness, $d_{Nb} \approx 4$–47 nm, prepared utilizing the wedge deposition technique described earlier. The $Cu_{1-x}Ni_x$ layer was flat with a thickness fixed at a physically infinite value of $d_{CuNi} = 56$ nm [13].

In the second kind of samples, the superconducting Nb layer was flat with a thickness fixed at a selected value in the range 6–15 nm. The deposition technique with moving magnetron described earlier provided high-quality niobium layers with superconducting T_{c0} of the stand-alone film as high as 5.5 K at $d_{Nb} \approx 5.7$ nm only. The ferromagnetic layer was wedge-shaped. A sketch of the layers stack is presented in the inset of Fig. 1.4a, and a transmission electron microscopy image of one of the samples is given in Fig. 1.4b. After cutting the final stack into strips across the $Cu_{1-x}Ni_x$ wedge gradient, a series of 36–40 samples were obtained with variable $Cu_{1-x}Ni_x$ layer thicknesses in the range $d_{CuNi} \approx 1$–35 nm, prepared at identical conditions in a single deposition run.

Rutherford backscattering spectrometry (RBS) has been used to evaluate the thickness of Nb and $Cu_{1-x}Ni_x$ layers as well as to check the composition of Cu and Ni in the deposited alloy layers (Fig. 1.4a). For details, see [13]. An advantage of RBS is that it is an absolute method that does not require standards for quantification. It allows to determine the thickness (via the areal density) of the layers with an accuracy of ±3% for $Cu_{1-x}Ni_x$ on the thick side of the $Cu_{1-x}Ni_x$ wedge, and ±5% for Nb and $Cu_{1-x}Ni_x$ on the thin side of the wedge. The Ni concentration in the $Cu_{1-x}Ni_x$ layer appeared to be almost constant ($x \approx 0.59$), showing a slight increase toward the thick side of the wedge. The thickness of the Nb layer is nearly constant along the uncut sample, $d_{Nb} \approx 7.8$ nm. Several samples were studied by TEM, a representative example is given in Fig. 1.4b.

The resistance measurements were performed by the DC four-probe method using a 10 μA sensing current in the temperature range 0.4–10 K when measuring with an Oxford Instruments "Heliox" ^{3}He cryostat, and a 2 μA sensing current in the range 40 mK–1.0 K when measuring in an Oxford Instruments dilution refrigerator "Kelvinox".

1.3 Results of Superconducting T_C Measurements and Discussion

The superconducting critical temperature, T_c, was determined from the midpoints of resistive transitions curves $R(T)$. The width of transition ($0.1 R_N$–$0.9 R_N$ criteria, where R_N is the normal state resistance just above T_c) for most of the investigated samples was below 0.2 K, thus allowing to determine the T_c with a good accuracy.

Figure 1.5a demonstrates the dependence of the superconducting transition temperature on the Nb layer thickness, $T_c(d_{Nb})$. It yields a critical thickness ($d_{Nb}^{cr} \approx 5.8$ nm) of the Nb layer down to which superconductivity survives in a metallic

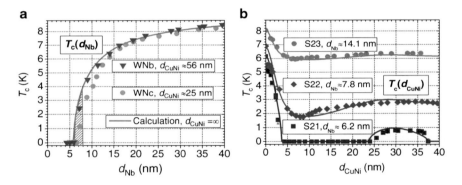

Fig. 1.5 Superconducting T_c as a function of the layer thickness. (**a**) Dependence of the superconducting transition temperature on the niobium layer thickness. Transition widths are within the point size if error bars not visible. The solid line is the result of calculations according to the theory [13, 26] for $d_{CuNi} \approx 56$ nm, the superconducting coherence length $\xi_S = 11.0$ nm, the ratio of the Sharvin conductances $N_F v_F / N_S v_S = 0.23$, the S/F-interface transparency parameter $T_F = 0.65$, $l_F/\xi_{F0} = 1.1$, $\xi_{F0} = 11.0$ nm. The calculated critical thickness is $d_{Nb}^{cr} = 5.8$ nm ($T_c \rightarrow 0$ K). The range of the Nb layer thickness most sensitive to d_{CuNi} variations is shaded in red. (**b**) Non monotonous $T_C(d_F)$ dependence for the Nb/Cu$_{1-x}$Ni$_x$ bilayers ($x = 0.59$). Solid curves are calculated with values of parameters as follows: (S23) $\xi_S = 10.0$ nm, $N_F v_F / N_S v_S = 0.22$, $T_F = 0.43$, $l_F/\xi_{F0} = 1.1$, $\xi_{F0} = 10.6$ nm; (S22) $\xi_S = 9.8$ nm, $N_F v_F / N_S v_S = 0.22$, $T_F = 0.55$, $l_F/\xi_{F0} = 1.1$, $\xi_{F0} = 10.6$ nm; (S21) $\xi_S = 9.6$ nm, $N_F v_F / N_S v_S = 0.22$, $T_F = 0.59$, $l_F/\xi_{F0} = 1.1$, $\xi_{F0} = 11.0$ nm. The BCS coherence length for Nb was always taken $\xi_{BCS} = 42$ nm. The calculations give no further reentrance of superconductivity for the S21 sample series above $d_{CuNi} > 40$ nm

contact with a bulk ferromagnet. The critical thickness is used to determine a constraint on two of the five physical parameters that enter the theory [26]. On the other hand, the $T_c(d_{Nb})$ measurements provide a range of the Nb layer thickness, within which superconductivity is most sensitive to variations of the magnetic layer thickness (the shaded area indicated in Fig. 1.5a). To observe the reentrant behavior of superconductivity, one should prepare samples with the Nb layer thickness in this range of $d_{Nb} \approx 6$–8 nm.

Figure 1.5b demonstrates the dependence of the superconducting transition temperature on the $Cu_{41}Ni_{59}$ layer thickness. The thickness of the flat Nb layer is fixed, $d_{Nb} \approx 14.1$ nm (S23 series), $d_{Nb} \approx 7.8$ nm (S22 series), and $d_{Nb} \approx 6.2$ nm (S21 series). The transition temperature, T_c, for the specimens with $d_{Nb} \approx 14.1$ nm reveals a nonmonotonous behavior with a shallow minimum at about $d_{CuNi} \approx 7.0$ nm. For the thinner niobium layer ($d_{Nb} \approx 7.8$ nm), the transition temperature shows a pronounced minimum with subsequent increase of T_c to above 2.5 K. For the thinnest Nb layer ($d_{CuNi} \approx 6.2$ nm), the superconducting T_c sharply drops upon increasing the ferromagnetic $Cu_{41}Ni_{59}$ layer thickness till a certain thickness $d_{CuNi} \approx 2.5$ nm. Then, in the range $d_{CuNi} \approx 2.5$–24 nm, the superconducting transition temperature vanishes (T_c is at least lower than the lowest temperature reached in our cryogenic setup, 40 mK). With a subsequent increase of the $Cu_{1-x}Ni_x$ layer thickness, superconductivity restores again at $d_{CuNi} \approx 25.5$ nm, reaching a level of about 0.8 K at $d_{CuNi} \approx 30$ nm, and then drops down again below 40 mK at $d_{CuNi} \approx 37.5$ nm. This phenomenon of a double suppression of superconductivity is the first experimental evidence for a multiple reentrant behavior of the superconducting state in S/F layered systems.

The data simulation procedure includes coordinated fitting of the $T_c(d_{Nb})$ and $T_c(d_{CuNi})$ dependences as shown in Fig. 1.5a, b respectively. The general fitting strategy is described in detail in our previous papers [13, 26]. The solid curves in the figures show results of the calculations for the "clean" case with parameters given in the figure caption. Although we used a common set of parameters at first, the superconducting coherence length, ξ_S, and the magnetic coherence length, ξ_{F0}, were varied within a 5% range, and the S/F interface transparency parameter, T_F, which generally lies in the range $[0, \infty)$, was varied within the range $[0.43, 0.65]$ to obtain better fits for the individual curves. These degrees of freedom that we allowed for the physical parameters are well within the scatter, which can be expected from variations of the deposition conditions from run to run. Calculations with the physical parameters of the S21 sample series, but for a slightly thicker Nb layer $d_{Nb} \approx 6.3$–6.4 nm, show that the next island of superconductivity is possible in the range $d_{CuNi} \approx 53$–70 nm with maximal T_c of about 0.3 K. We will search for the second reentrance of superconductivity in our further studies.

1.4 Conclusions

To conclude, we report on the experimental observation of the reentrant behavior of superconductivity and a unique double suppression of superconductivity in S/F bilayers. As S material, Nb with constant layer thickness (≈ 6.2 nm) was used, and as F material $Cu_{1-x}Ni_x$ alloy ($x \approx 0.59$) with variable layer thickness. The experimental realization of the reentrant superconductivity phenomenon is an essential progress toward the fabrication of a $F_1/S/F_2$ superconducting spin switch [27–30] for superconducting spintronics.

Acknowledgements The authors are grateful to J. Aarts, C. Attanasio, A.I. Buzdin, M.Yu. Kupriyanov, V. Oboznov, S. Prischepa, and Z. Radovic for stimulating discussions, to J. Lindner, J. Moosburger-Will, and W. Reiber for assistance in TEM sample preparation and measurements. The work was partially supported by DFG through SFB-484, BMBF (project No MDA01/007), RFBR (projects No 07-02-00963, No 08-02-90105-Mol_a, 09-02-12176-ofi_m, 09-02-12260-ofi_m and No 08.820.05.28RF) and the Program of RAS "Spintronics".

References

1. P.G. De Gennes, E. Guyon, Phys. Lett. **3**, 168–169 (1963)
2. N.R. Werthamer, Phys. Rev. **132**, 2440–2445 (1963)
3. Z. Radović, L. Dobrosavljević-Grujić, A.I. Buzdin, J. Clem, Phys. Rev. B **38**, 2388–2393 (1988)
4. E.A. Demler, G.B. Arnold, M.R. Beasley, Phys. Rev. B **55**, 15174–15182 (1997)
5. A.A. Golubov, M.Yu. Kupriyanov, E. Il'ichev, Rev. Mod. Phys. **76**, 411–469 (2004)
6. I.F. Lyuksyutov, V.L. Pokrovsky, Adv. Phys. **54**, 67–136 (2005)
7. A.I. Buzdin, Rev. Mod. Phys. **77**, 935–976 (2005)
8. F.S. Bergeret, A.F. Volkov, K.B. Efetov, Rev. Mod. Phys. **77**, 1321–1373 (2005)
9. J.S. Jiang, D. Davidović, D.H. Reich, C.L. Chien, Phys. Rev. Lett. **74**, 314–317 (1995)
10. Th. Mühge, N.N. Garif'yanov, Yu.V. Goryunov, G.G. Khaliullin, L.R. Tagirov, K. Westerholt, I.A. Garifullin, H. Zabel, Phys. Rev. Lett. **77**, 1857–1860 (1996)
11. L.V. Mercaldo, C. Attanasio, C. Coccorese, L. Maritato, S.L. Prischepa, M. Salvato, Phys. Rev. B **53**, 14040–14042 (1996)
12. I.A. Garifullin, D.A. Tikhonov, N.N. Garif'yanov, L. Lazar, Yu.V. Goryunov, S.Ya. Khlebnikov, L.R. Tagirov, K. Westerholt, H. Zabel, Phys. Rev. B **66**, 020505 (2002)
13. V.I. Zdravkov, A.S. Sidorenko, G. Obermeier, S. Gsell, M. Schreck, C. Müller, S. Horn, R. Tidecks, L.R. Tagirov, Phys. Rev. Lett. **97**, 057004 (2006)
14. V.V. Ryazanov, V.A. Oboznov, A.Yu. Rusanov, A.V. Veretennikov, A.A. Golubov, J. Aarts, Phys. Rev. Lett. **86**, 2427–2430 (2001)
15. T. Kontos, M. Aprili, J. Lesueur, X. Grison, Phys. Rev. Lett. **86**, 304–307 (2001)
16. P. Fulde, R. Ferrell, Phys. Rev. **135**, A550–A563 (1964)
17. A.I. Larkin, Yu.N. Ovchinnikov, Zh. Eksp. Teor. Fiz. **47**, 1136–1146 (1964); [Sov. Phys. JETP **20**, 762–769 (1965)]
18. J. Aarts, J.M.E. Geers, E. Brück, A.A. Golubov, R. Coehoorn, Phys. Rev. B **56**, 2779–2787 (1997)
19. L.R. Tagirov, Physica C **307**, 145–163 (1998)
20. M. Born, E. Volf, *Principles of Optics*, 4th edn (Pergamon Press, New York, 1968), Chapter 7
21. T. Tokuyasu, J.A. Sauls, D. Rainer, Phys. Rev. B **38**, 8823–8832 (1988)
22. A. Cottet, W. Belzig, Phys. Rev. B **72**, 180503(R) (2005)
23. A. Cottet, Phys. Rev. B **76**, 224505 (2007)
24. M.G. Khusainov, Yu.N. Proshin, Phys. Rev. B **56**, 14283–14286 (1997); Erratum: Phys. Rev. B **62**, 6832–6833 (2000)
25. B.P. Vodopyanov, L.R. Tagirov, Pis'ma v ZhETF **78**, 1043–1047 (2003); [JETP Lett **78**, 555–559 (2003)]

26. A.S. Sidorenko, V.I. Zdravkov, A. Prepelitsa, C. Helbig, Y. Luo, S. Gsell, M. Schreck, S. Klimm, S. Horn, L.R. Tagirov, R. Tidecks, Ann. Phys. **12**, 37–50 (2003)
27. P.G. De Gennes, Phys. Lett. **23**, 10–11 (1966)
28. G. Deutscher, F. Meunier, Phys. Rev. Lett. **22**, 395–396 (1969)
29. L.R. Tagirov, Phys. Rev. Lett. **83**, 2058–2061 (1999)
30. A.I. Buzdin, A.V. Vedyayev, N.V. Ryzhanova, Europhys. Lett. **48**, 686–691 (1999)

Chapter 2
Aharonov–Bohm Oscillations in Small Diameter Bi Nanowires

L. Konopko

Abstract The Aharonov–Bohm effect (AB) exists in cylindrical wires as the magnetoresistance (MR) oscillations with a period ΔB that is proportional to Φ_0/S, where $\Phi_0 = h/e$ is the flux quantum and S is the wire cross section. The AB-type longitudinal MR oscillations with period $\Delta B = \Phi_0/S$ caused by electrons undergoing continuous grazing incidence at the wire wall have been observed previously at 4.2 K in single bismuth nanowires with a diameter $0.2 < d < 0.8\,\mu\text{m}$ grown by the Ulitovsky technique. We present here our results of the observation of AB oscillations with period $\Delta B = h/e$ and $\Delta B = h/2e$ on single Bi nanowires with a diameter $d = 45$–73 nm. The single nanowire samples were prepared by improved Ulitovsky technique and represented cylindrical single crystals with (10$\bar{1}$1) orientation along the wire axis. Due to very low effective masses of electrons and holes, electronic quantum confinement effects induce a semimetal-to-semiconductor transformation (SMSC) for wires with diameters below 50 nm. Our estimation of thermal energy gap from $R(T)$ dependence for 50 nm Bi wire gives the value of 14 meV. The surface of Bi nanowire supports surface states, with carrier densities of around $5 \times 10^{12}\,\text{cm}^{-2}$ with strong spin-orbit interactions. From $B \sim 8$ T down to $B = 0$, the extremums of h/2e oscillations are shifted up to 3π at $B = 0$, which is the manifestation of Berry phase shift. We connect the existence of $h/2e$ oscillations with weak localizations on surface states of Bi nanowires according to the Altshuller–Aronov–Spivak theory.

2.1 Introduction

Nanowire systems have been investigated for many decades. At present, they have become the focus of intense experimental and theoretical investigation due to their scientific and technologic interest. The most exciting opportunity is that of an ideal quantum wire of a diameter d that is less than the Fermi wavelength and with the Fermi level chosen such that the nanowire transport is controlled by a single conduction channel. The properties of this one-dimensional system or quantum wire have been investigated theoretically, and the case of Bi nanowires was studied by Hicks and Dresselhaus [1].

Bismuth is a particularly favorable material to study the electronic properties of quantum wires due to its small electron effective mass and high carrier mobility [2]. De Haas–van Alphen and Shubnikov–de Haas (SdH) effects [3] provide an unambiguous measure of the charge density and the anisotropy of the Fermi surface (FS) of bulk crystalline Bi [4], which consists of three electron pockets at the L point and a T-point hole pocket. SdH oscillations also have been studied in single-crystal films of Bi [5]. The effective band overlap energy, E_0, and the Fermi energy, E_F, are 37 and 26 meV respectively, levels which result in small electron and hole densities ($n_i = p_i = 3 \times 10^{17}$ cm^{-3} at a temperature of 4 K). Quantum confinement effects, which decrease E_0, become relevant to quantum wires with a diameter $d \approx 2\hbar/\sqrt{2m^*E_0}$, where m^* is the corresponding electron in-plane effective mass transverse to the wire axis. For wires oriented along C_3 (the trigonal direction), $m^* = 0.0023$ and the relevant diameter is 42 nm. Detailed calculations [2] show that a semimetal-to-semiconductor (SMSC) transition occurs for $d_c \sim 55$ nm for wires oriented along the trigonal direction. Various experimental results support this theory.

It is well known that quantum interference effects are present in superconducting devices and in very small pure metallic rings and cylinders. In particular, in the presence of magnetic flux Aharonov–Bohm (AB) oscillations [6] may occur in doubly connected systems [7–9]. For a normal metal, the period of these oscillations is $\Phi_0 = h/e$ (the flux quantum). Such effects should vanish once the elastic mean free path of the electrons is smaller than the size of the system. There are two types of quantum interference effects in normal conductors of long carrier's mean free path that cause a magnetoresistance (MR) oscillation with a period ΔB that is proportional to Φ_0/S, where S is the wire cross section. The first effect is a Dingle oscillation that results from the quantization of the electron energy spectrum. The second type of oscillation, with a period $\Delta B = \Phi_0/S$, is caused by electrons undergoing continuous grazing incidence at the wire walls. These are termed "whispering gallery" modes analogous to the acoustical phenomenon. This type of longitudinal MR oscillations have been observed at 4.2 K in single nanowires grown by Ulitovsky technique with a diameter $0.2 < d < 0.8$ μm [10–12] and in a Bi nanowire arrays 270 nm in diameter [13]. For the disordered cylindrical samples with short mean free path (compared with the circumference of the cylinder) the new type of AB oscillations with a period $\Delta B = \Phi_0/2S$ was predicted by Al'tshuler, Aronov, and Spivak (AAS) [14]. This effect arises from the interference of pairs of coherent electron waves circumscribing the cylinder. These oscillations were observed by Sharvin and Sharvin [15] on the Mg cylinder 1 μm in diameter and 1 cm long.

Since the introduction of the AB effect, the phase factor has been studied intensively. Berry [16] showed that, even in the absence of electromagnetic fields, when a quantum state undergoes an adiabatic evolution along a closed curve in parameter space, it develops a phase that depends only on this curve. To observe Berry's phase in an electronic system with spin, Loss et al. [17] proposed to study transport in a mesoscopic ring structure in the presence of an orientationally inhomogeneous magnetic field. This can be experimentally implemented via fabricating the ring from a material with inversion asymmetry and spin–orbit (SO) interaction.

In this paper, we present our results of the observation of AB oscillations with periods $\Delta B = \Phi_0/S$ and $\Delta B = \Phi_0/2S$ on single Bi nanowires with a diameter $45 < d < 75$ nm. The manifestation of Berry phase shift for $h/2e$ oscillations will be discussed.

2.2 Experimental

The Bi nanowires were fabricated using the Ulitovsky technique, by which a high-frequency induction coil melts a 99.999%-pure Bi boule within a borosilicate glass capsule, simultaneously softening the glass. Glass capillaries containing Bi filament were produced by drawing material from the glass [10, 11, 18]. Schematic diagram of the Ulitovsky fabrication process of the Bi nanowire is shown in Fig. 2.1b. Encapsulation of the Bi filament in glass protects it from oxidation and mechanical stress. It has been observed that individual nanowires are single crystals, the crystalline structure of which is determined by Laue x-ray diffraction and SdH methods. The Bi in the microwires can be viewed as cylindrical single crystals with the (10$\bar{1}$1) orientation along the wire axis. In this orientation, the wire axis makes an angle of

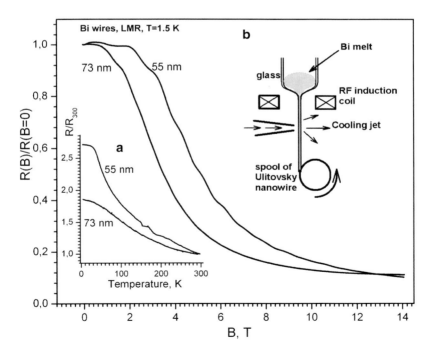

Fig. 2.1 Magnetic field dependence of the longitudinal MR for a 55 and 73 nm Bi nanowires, $T = 1.5$ K. (**a**) Temperature dependences of the resistance for a 55 and 73 nm Bi nanowires and (**b**) schematic diagram of the Ulitovsky fabrication process of the Bi nanowire

19.5° with the bisector axis C_3 in the bisector-trigonal plane also; the trigonal axis C_3 is inclined to the wire axis at an angle of 70°; and one of the binary axes C_2 is perpendicular to it. Electrical connections to the nanowires were performed using $In_{0.5}Ga_{0.5}$ eutectic. This type of solder consistently makes good contacts, compared with other low-melting-point solders, but it has the disadvantage that it diffuses at room temperature into the Bi nanowire rather quickly. Consequently, the nanowire has to be used in the low-temperature experiment immediately after the solder is applied. The samples used in this work are, to date, the smallest diameter single-Bi wires for which the electronic transport at low temperatures has been reported.

All measurements were performed at the High Magnetic Field Laboratory (Wroclaw, Poland) in superconducting solenoid in magnetic fields of up to 14 T at temperatures 1.5–4.2 K. The samples were measured in a two-axis rotator, in which the sample holder was rotated by means of a worm drive and a step motor. Together with MR data, we registered the first derivative of the MR using modulation technique (small AC magnetic field $\Delta B = 7.5 \times 10^{-5}$ T was produced by special modulation superconducting coil at frequency 13.7 Hz). A 7265 DSP Lock-in Amplifier was used for the separation of derivative signal.

2.3 Results and Discussion

Figure 2.1 shows the magnetic field dependence of the longitudinal MR for Bi nanowires with $d = 55$ and 73 nm. $R(B)$ decreases for increasing magnetic field and it is typical of Bi nanowires of large diameter [18]. This effect has been observed in many studies and by many groups in almost all samples of Bi nanowires, even those of small diameter [2,19]. This phenomenon is a Chambers effect, which occurs when the magnetic field focuses electrons toward the core of the wire (away from the surface), thereby avoiding surface collisions. We have observed the nonmonotonic changes of longitudinal MR that are equidistant in the magnetic field. Fig. 2.1a presents the temperature dependence of resistance R_T/R_{300} of Bi nanowires with $d = 55$ and 73 nm. The $R(T)$ dependencies have "semiconductor" character, i.e., the resistance grows in the whole range of temperatures. For $T > 100$ K, the nanowires' resistance $R(T) \sim \exp(\Delta/2k_B T)$. Δ is found to be 10 ± 5 meV for both the 55 and 73 nm wires. Following Choi et al. [20], Δ is interpreted as the energy gap between the electron and hole band in the core of the nanowires. The values of Δ that are observed are in rough agreement with the theoretical work [2], which indicates that the band overlap decreases substantially below the value for bulk Bi (38 meV) because of quantum confinement. Therefore, one expects the electron and hole densities in our nanowires to be less than in bulk Bi. It can be surmised that the low-temperature electronic transport that is observed is mediated by surface states.

Figure 2.2 shows the oscillation part of magnetic field derivative of the longitudinal MR of 55 nm nanowire. Part (a) shows fast Fourier transform (FFT) spectra of this oscillation. Longitudinal MR oscillations that are equidistant in the magnetic field and decrease in amplitude have been observed for the first time in magnetic fields of up to 14 T in Bi single-crystal nanowires with $d < 80$ nm

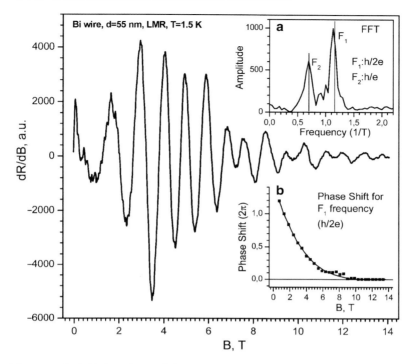

Fig. 2.2 Magnetic field dependence of derivative of longitudinal MR for a 55 nm Bi nanowires, $T = 1.5$ K (the monotonic part is subtracted). (**a**) FFT spectra of the oscillations and (**b**) changes of maxima positions versus B for $h/2e$ oscillations which were converted into the values of phase shift of high field harmonic oscillation

at $T = 1.4 \div 4.2$ K [21, 22]. In contrast to oscillations that have been observed in thick Bi microwires ($0.2 < d < 0.8\,\mu$m, $\Delta B_1 = h/e$ and $\Delta B_2 = 1.4 h/2e$), they exist in a wide range of magnetic field (up to 14 T) and have two periods: $\Delta B_1 = h/e$ and $\Delta B_2 = h/2e$. Finding the $\tilde{h}/e$ and the $\tilde{h}/2e$ modes together is not surprising; this been observed in rings where it has been interpreted to be caused by the interference of electrons that encircle the ring twice [23]. Due to SMSC in our nanowires carrier concentration in wires core very small, it is unlikely that they excite this oscillations. Angle-resolved photoemission spectroscopy (ARPES) studies of planar Bi surfaces have shown that they support surface states, with carrier densities Σ of around 5×10^{12} cm^{-2} and large effective masses m_Σ of around 0.3 [24]. The observed effects are consistent with theories of the surface of nonmagnetic conductors whereby Rashba SO interaction gives rise to a significant population of surface carriers [25]. Measurements of the FS of small-diameter bismuth nanowires employing the SdH method [26] have also been interpreted by assuming the presence of surface charge carriers with $\Sigma = 2 \times 10^{12}$ cm^{-2} and with the same effective mass as in ARPES.

Given the bulk electron n and hole p densities (in undoped Bi, $n = p = 3 \times 10^{17}$ cm^{-3} at 4 K) and the surface density Σ measured by ARPES or by SdH, one expects the surface carriers to become a clear majority in nanowires with diameters below 100 nm at low temperatures; the ratio of surface carrier density to bulk electrons or holes is 12 for 55 nm wires. At that point, the nanowire should effectively become a conducting tube. The electrical transport properties of nanotubes are unique, because the wavelike nature of the charge carriers manifests itself as a periodic oscillation in the electrical resistance as a function of the enclosed magnetic flux $\Phi = (\pi/4) d^2 B$.

Using FFT analysis, we have separated every frequency for longitudinal MR of 55 nm nanowire. As it has appeared h/e oscillation is harmonic but the extrema of $h/2e$ oscillation lie on a straight line only for $B > 8$ T and deviate step-by-step from it at low magnetic fields. After converting the low field extrema positions to the phase shift of high-field harmonic oscillations, the phase shift curve (Fig. 2.2b) was obtained. Mathematic simulation was used for testing this method. From $B \sim 8$ T down to $B = 0$, the extrema of $h/2e$ oscillations are shifted up to 3π at $B = 0$, which is the manifestation of Berry phase shift due to electron moving in a nonuniform magnetic field.

The derivative of MR for 55 nm bismuth nanowire was measured at various inclined angels of up to 90° of magnetic fields relative to nanowire's axis in two mutually perpendicular planes (Fig. 2.3 shows the FFT spectra for one of this plane). The observed angle variation of the periods is not in the agreement with the theoretical dependence $\Delta(\alpha) = \Delta(0)/\cos\alpha$ of the size effect oscillations of the "flux

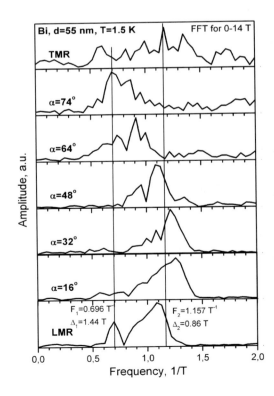

Fig. 2.3 Dependence of the FFT spectra for a 55 nm Bi nanowires, $T = 1.5$ K on the angle α between the direction of the applied magnetic field and the wire axis; $\alpha = 0$ corresponds to the longitudinal MR and $\alpha = 90°$ – to the transverse MR

quantization" type when the period of oscillations in magnetic field depends only on the component B_X, parallel to the axis of the cylindrical sample. According to the experimental data, the shifts of oscillation frequencies at the same angles depend on the plane of sample rotation. The derivative of MR was measured at various rotational angles of nanowire when the axis of the wire was mounted perpendicularly to the magnetic field. Even in this case, the equidistant oscillations of MR exist in transverse magnetic fields under certain rotation angles. This means that the nature of oscillations is connected with Bi FSs and is very complicated. We connect the existence of $h/2e$ oscillations with weak localizations on surface states of Bi nanowires according to the AAS theory.

2.4 Conclusions

To conclude, the AB oscillations in single-crystal Bi nanowires with glass coating with diameter from 45 to 75 nm were studied. The R(T) dependencies for Bi nanowires have "semiconductor" characteristics. For $T > 100$ K, the nanowires' resistance, $R(T) \sim \exp(\Delta/2k_B T)$. Δ is the energy gap between the electron and hole band in the core of the nanowires; Δ is found to be 10 ± 5 meV for both the 55 and 73 nm wires. This means that semimetal to semiconductor transformation exist in our nanowires. ARPES studies of planar Bi surfaces have shown that due to Rashba spin-orbit interaction they support surface states, with carrier densities Σ of around 5×10^{12} cm^{-2} and large effective masses m_Σ of around 0.3. Taking into account these properties, the Bi nanowire should effectively become a conducting tube. The AB oscillations with period $\Delta B = h/e$ and $\Delta B = h/2e$ were observed. According to FFT from $B \sim 8$ T down to $B = 0$, the extrema of $h/2e$ oscillations are shifted up to 3π at $B = 0$, which is the manifestation of Berry phase shift due to electron moving in a nonuniform magnetic field. We connect the existence of $h/2e$ oscillations with weak localizations on surface states of Bi nanowires according to the AAS theory. The observed angle variation of the periods is not in agreement with the theoretical dependence $\Delta(\alpha) = \Delta(0)/\cos\alpha$ of the size effect oscillations of the "flux quantization" type. Moreover, the equidistant oscillations of MR exist in transverse magnetic fields under certain rotation angles. This means that the nature of oscillations is connected with Bi FSs and is very complicated.

Acknowledgments This work was supported by the Academy of Sciences of Moldova Grant No 08.805.05.06A.

References

1. L.D. Hicks, M.S. Dresselhaus, Phys. Rev. B **47**, 16631 (1993)
2. Y. Lin, X. Sun, M.S. Dresselhaus, Phys. Rev. B **62**, 4610 (2000)
3. V.S. Edel'man, Adv. Phys. **25**, 555 (1976)

4. G.E. Smith, G.A. Baraf, J.M. Rowell, Phys. Rev. **135**, A1118 (1964)
5. F.Y. Yang et al., Phys. Rev. B **61**, 6631 (2000)
6. Y. Aronov, D. Bohm, Phys. Rev. **115**, 485 (1959)
7. A. Shablo, T. Narbut, S. Tyurin, I. Dmitrenko, Pis'ma Zh. Exp. Teor. Fiz. **13**, 457 (1974); JETP Lett. **19**, 146 (1974)
8. R. Dingle, Proc. R. Soc. Lond. A **47**, 212 (1952)
9. I.O. Kulik, Pis'ma Zh. Exp. Teor. Fiz. **5**, 423 (1967); JETP Lett. **5**, 145 (1967)
10. N.B. Brandt, D.V. Gitsu, A.A. Nikolaeva, Ya.G. Ponomarev, Zh. Exp. Teor. Fiz. **72**, 2332–2334 (1977); Sov. Phys. JETP **45**, 1226 (1977)
11. N.B. Brandt, E.N. Bogachek, D.V. Gitsu, G.A. Gogadze, I.O. Kulik, A.A. Nikolaeva, Ya.G. Ponomarev, Fiz. Nizk. Temp. **8**, 718 (1982); Sov. J. Low Temp. Phys. **8**, 358 (1982)
12. N.B. Brandt, D.V. Gitsu, V.A. Dolma, Ya.G. Ponomarev, Zh. Exp. Teor. Fiz. **92**, 913 (1987)
13. T.E. Huber, K. Celestine, M.J. Graf, Phys. Rev. B **67**, 245317 (2003)
14. B.L. Al'tshuler, A.G. Aronov, B.Z. Spivak, Zh. Exp. Teor. Fiz., Pis. Red. **33**, 101 (1981); JETP Lett. **33**, 94 (1981)
15. D.Yu. Sharvin, Yu.V. Sharvin, Zh. Exp. Teor. Fiz., Pis. Red. **34**, 285 (1981); JETP Lett. **34**, 272 (1981)
16. M.V. Berry, Proc. R. Soc. Lond. A **45**, 392 (1984)
17. D. Loss, P.M. Goldbart, A.V. Balatsky, Phys. Rev. Lett. **65**, 1655 (1990)
18. D. Gitsu, L. Konopko, A. Nikolaeva, T.E. Huber, Appl. Phys. Lett. **86**, 102105 (2005)
19. J. Heremans, C.M. Thrush, Y.M. Lin, S. Cronin, Z. Zhang, M.S. Dresselhaus, J.F. Mansfield, Phys. Rev. B **61**, 2921 (2000)
20. D.S. Choi, A.A. Balandin, M.S. Leung, G.W. Stupian, N. Presser, S.W. Chung, J.R. Heath, A. Khitun, K.L. Wang, Appl. Phys. Lett. **89**, 141503 (2006)
21. D. Gitsu, T. Huber, L. Konopko, A. Nikolaeva, AIP Conf. Proc. **850**, 1409 (2006)
22. A. Nikolaeva, D. Gitsu, L. Konopko, M.J. Graf, T.E. Huber, Phys. Rev. B **77**, 075332 (2008)
23. V. Chandrasekhar, M.J. Rooks, S. Wind, D.E. Prober, Phys. Rev. Lett. **55**, 1610 (1985)
24. Yu.M. Koroteev, G. Bihlmayer, J.E. Gayone, E.V. Chulkov, S. Blugel, P.M. Echenique, Ph. Hofmann, Phys. Rev. Lett. **93**, 046403 (2004)
25. Ph. Hofmann, Prog. Surf. Sci. **81**, 191 (2006)
26. T.E. Huber, A. Nikolaeva, D. Gitsu, L. Konopko, C.A. Foss Jr., M.J. Graf, Appl. Phys. Lett. **84**, 1326 (2004)

Chapter 3
Point-Contact Study of the Superconducting Gap in the Magnetic Rare-Earth Nickel-Borocarbide RNi$_2$B$_2$C (R = Dy, Ho, Er, Tm) Compounds

Yu.G. Naidyuk, N.L. Bobrov, V.N. Chernobay, S.-L. Drechsler, G. Fuchs, O.E. Kvitnitskaya, D.G. Naugle, K.D.D. Rathnayaka, L.V. Tyutrina, and I.K. Yanson

Abstract We present an overview of the efforts in point-contact (PC) study of the superconducting (SC) gap in the antiferromagnetic (AF) nickel-borocarbide compounds RNi$_2$B$_2$C (R = Dy, Ho, Er, Tm), for which the energy scales of AF and SC order, measured by the Neel temperature T_N and the critical temperature T_c, respectively, can be varied over a wide range. The SC gap was determined from the experimental dV/dI curves of PCs employing the well-known BTK theory of conductivity for normal metal-superconductor PCs accounting Andreev reflection. Additionally, the mentioned theory including pair-breaking effect due to magnetic impurities was employed and a multiband structure of the title compounds was taken into consideration. Recent directional PC study of the SC gaps gives evidence for the anisotropic two-band (two-gap) nature of SC-ty in R = Er ($T_N \approx 6$ K $< T_c \approx 11$ K). Additionally, a distinct decrease of both gaps in this compound in the AF state is observed. The SC gap in R = Ho ($T_N \approx 5.2$ K $< T_c \approx 8.5$ K) exhibits below $T^* \approx 5.6$ K a standard single-band BCS-like dependence vanishing above T^*, where a specific magnetic ordering starts to play a role. For R = Tm ($T_N \approx 1.5$ K $< T_c \approx 10.5$ K) a decrease of the SC gap is observed below ∼5 K, while for R = Dy ($T_N \approx 10.5$ K $> T_c \approx 6.5$ K) the SC gap has BCS-like dependence in the AF state. Distinct features of the SC gap behavior in the mentioned magnetic superconductors are discussed.

3.1 Introduction

The RNi$_2$B$_2$C (R denotes mainly a rare earth element) family of ternary superconductors is very attractive for studies of fundamental questions in superconductivity and its interplay with magnetic order [1]. This family attracts attention because the superconducting (SC) critical temperature in RNi$_2$B$_2$C is relatively high, and their SC properties exhibit often unconventional behavior and superconductivity and magnetic order compete in some of these materials. As it was shown in [2], to describe properly the SC properties of nickel borocarbides a multiband scenario is required. The manifestation of the mentioned extraordinary properties of RNi$_2$B$_2$C is intimately dependent on the chemical composition and crystal perfectness Therefore continuous progress in synthesis of highquality single crystal samples leads to a deeper understanding of their fundamental physics. This concerns the nature of Cooper pairing and attractive interaction along with competition of SC and magnetic ordered states.

Studies of the directional, temperature and magnetic field dependence of the SC gap can provide an insight into the SC ground state of RNi$_2$B$_2$C. This can be done in the most direct way by point-contact (PC) [3], scanning tunneling or photoemission spectroscopy. The last method has so far a lower resolution The tunneling spectroscopy is extremely sensitive to the surface or even upper layers condition. Therefore, nickel borocarbides are to date mostly studied by PC spectroscopy. Thus, in this work we overview the efforts in PC studies of the SC gap in antiferromagnetic (AF) nickel-borocarbide superconductors RNi$_2$B$_2$C(R = Dy, Ho, and Er, Tm) based on published papers [4–15].

3.2 Experimental

Here we will concentrate on results measured on the best single crystals reported so far. HoNi$_2$B$_2$C ($T_N \approx 5.2$ K $< T_c \approx 8.5$ K) and TmNi$_2$B$_2$C ($T_N \approx 1.5$ K $< T_c \approx 10.5$ K) crystals were grown by a floating zone technique with optical heating at IFW Dresden, while ErNi$_2$B$_2$C ($T_N \approx 6$ K $< T_c \approx 11$ K) and DyNi$_2$B$_2$C ($T_N \approx 10.5$ K $> T_c \approx 6.5$ K) were grown at Ames Laboratory by a Ni$_2$B high-temperature flux growth method. PCs were established at the liquidhelium temperature by standard "needle-anvil" methods [3], using as a "needle" a small sharpened bar or thin wire of a noble metal like Cu or Ag. The "anvil" or sample surface was prepared by chemical etching or by cleavage as reported in corresponding papers [4–15]. The SC gap was evaluated from the measured $dV/dI(V)$ dependences of PCs using both the standard BTK theory [16] and a similar theory (in the case of Er), which includes the pairbreaking effect of magnetic impurities [17].[1] This is important for the mentioned borocarbides because of the presence of magnetic moments in rare earth ions.

3.3 Results and Discussion

The SC gap Δ_0, results in the appearance of pronounced minima symmetrically placed at $V \approx \pm\Delta/e$ on the dV/dI characteristic of a normal metal-superconductor contact at $T \ll T_c$ (see inset in Fig. 3.1). The measured dV/dI curves of SC nickel

---

[1] The theory [17] was used to fit dV/dI for ErNi$_2$B$_2$C PCs. In this case there is a difference between the SC gap Δ_0, and the SC order parameter, Δ, namely, $\Delta = \Delta\left(1 - \gamma^{2/3}\right)^{3/2}$, where γ is the pair-breaking parameter. However, for uniformity we used definition "one-gap" or "two-gap" fit, although in the case of ErNi$_2$B$_2$C definition "two-OP" fit is proper.

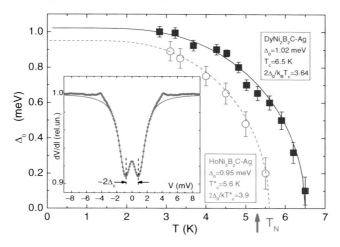

Fig. 3.1 Gap behavior in HoNi$_2$B$_2$C (*circles*) [11, 12] and DyNi$_2$B$_2$C (*squares*) [9] established by a BTK fit of PC dV/dI curves. The *arrow* shows the Neel temperature for the Ho compound, while T_N in Dy is about 10.5 K, i.e. out of the x-axis. *Inset*: example of a dV/dI curve (*symbols*) fitted by BTK theory [16] (*thin curve*)

borocarbides exhibit one pair of minima as in the case of a singlegap superconductors [3] Therefore a single-gap approach is usually used to fit experimental data. It is seen from the inset in Fig. 3.1 that the one-gap fit reasonably describes dV/dI for $R =$ Ho as well as for $R =$ Dy [9] (not shown). The obtained $\Delta_0(T)$ in Fig. 3.1 has a BCS-like temperature dependence in both cases However, $\Delta_0(T)$ vanishes at $T^* \approx 5.6$ K for $R =$ Ho, well below the bulk critical temperature of $T_c \approx 8.5$ K. It was suggested in [12] that superconductivity in the commensurate AF phase in the $R =$ Ho compound survives at a special nearly isotropic Fermi surface sheet, while the gap suppression at T^* may be caused by a peculiar magnetic order developing in this compound above the AF state at $T_N \approx 5.2$ K. Note, that the SC state in HoNi$_2$B$_2$C in this region is sensitive to the current density j (see Fig. 3.5 in [18]). Thus, for $j \approx 100$ A cm^{-2} a decrease of T_c by about 1 K, a remarkable broadening of the SC transition and an enhanced reentrant behavior at T_N were found. Note that the current density in PC is several order of magnitude higher [3] than that used in [18], which could be the reason for the suppression of the SC gap above T^* in the region of "exotic" magnetically ordered states.

Recently, an interesting peculiar behavior of $\Delta_0(T)$ in TmNi$_2$B$_2$C (Fig. 3.2) was found [13]. The SC gap has a maximum around 4–5 K and further decreases for decreasing temperature. This is in line with the behavior of the upper critical field along the c-axis. Apparently, AF Fluctuations occurring above the magnetically ordered state at $T_N = 1.5$ K are responsible for the decrease of the SC gap observed at low temperatures.

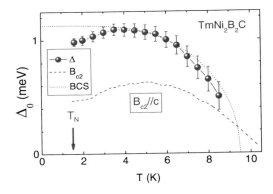

Fig. 3.2 Gap behavior in TmNi$_2$B$_2$C established by a BTK fit of PC dV/dI curves. The *dashed curve* shows the upper critical field, B_{c2}, in TmNi$_2$B$_2$C for B along the c-axis [13]. The scale of B_{c2} has been omitted for the sake of simplicity

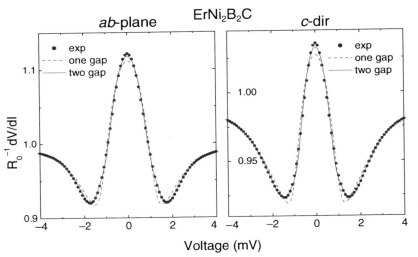

Fig. 3.3 Example of the BTK fit of the PC dV/dI curves for ErNi$_2$B$_2$C using the "one-gap" and "two-gap" approaches [14]

As it was shown in [14], the "one-gap" approach to fit the measured highquality dV/dI curves[2] for ErNi$_2$B$_2$C results in a visible discrepancy between the fit and the data at the minima position and at zero bias (Fig. 3.3). At the same time, a "two-gap" approach allows a better fit of the experimental curves for ErNi$_2$B$_2$C (Fig. 3.3). As was mentioned earlier, the upper critical field $H_{c2}(T)$ of nonmagnetic borocarbides $R = $ Y and Lu [2] can be properly described only by a two-band model. Therefore, the finding of two SC gaps (with the about the same values) in magnetic ErNi$_2$B$_2$C from one side testifies to the similarities in the electronic band

---

[2] High quality means that dV/dI has pronounced minima and additionally the absence of spikelike structures above the minima (see, for example, the inset of ** Fig. 3.1) and other irregularities that often accompany the measured dV/dI.

3 Point-Contact Study of the Superconducting Gap

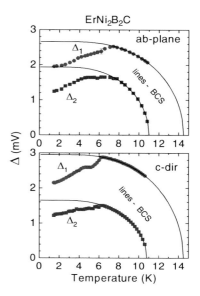

Fig. 3.4 Temperature dependence of the larger OP (*circles*) and the smaller OP (*squares*) for ErNi$_2$B$_2$C for the two main crystallographic directions according to [14]

structure of mentioned compounds, and from the other side it points to the fact that superconductivity and magnetism develop also in ErNi$_2$B$_2$C in different bands.

Let us turn to the temperature dependence of the SC order parameter[3] (OP) in ErNi$_2$B$_2$C shown in Fig. 3.4. It is clearly seen that both OPs start to decrease near or below the Neel temperature. A similar decrease of the SC gap in ErNi$_2$B$_2$C in the AF state was reported by STM measurements in [19] and recently, by laser-photoemission spectroscopy [20]. Such a gap decrease in the AF state is also in line with the Machida theory [21] in which a spin density wave ordering competes with superconductivity.

On the other hand, in the paramagnetic state $\Delta_0(T)$ in ErNi$_2$B$_2$C is close to the BCS-like behavior, and only the abrupt vanishing of the larger OP at T_c is rather unexpected. It turned out [14] that the contribution of the larger OP to dV/dI is also temperature dependent decreasing with increasing temperature and containing only about 10% close to T_c. Therefore, it seems that the larger OP disappears at T_c due to a "shrinking" of the corresponding SC part of the Fermi surface.

It should be noted that the AF structure in Er and Tm nickel borocarbides is an incommensurate spin density wave. As it is seen from Fig. 3.2, $\Delta_0(T)$ for TmNi$_2$B$_2$C deviates from the BCS behavior while approaching T_N. Contrarily, Ho and Dy compounds with commensurate AF order display a BCS-like gap. Of

---

[3] In the case of pair-breaking by magnetic impurities gapless superconductivity may occur and the SC ** order parameter (OP) cannot be measured by a gap as in the case of usual superconductors. The OP is generally measured by the condensate density of Cooper pairs. For the relation between the OP and the derivative of the density of states measured in PC studies in this case and the shape of the dV/dI curves see [17].

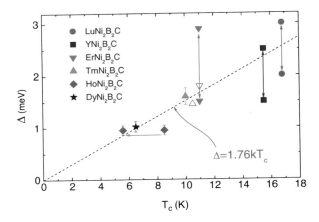

Fig. 3.5 The SC gap or OP (solid symbols) established by a PC study as a function of the critical temperature T_c in RNi$_2$B$_2$C including also the nonmagnetic compounds for R = Y and Lu. For R = Y, the extremes of the anisotropic SC gap are presented, while for R = Lu and Er the small and large OPs are shown. For R = Ho, T_c is shifted to T^* = 5.6 K (see text for explanation). The open symbols show the SC gap determined from tunneling spectroscopy [22, 23]. The BCS ratio is shown by the dashed straight line

course, in the case of R = Tm measurements below $T_N \approx 1.5$ K are very desirable to trace similarity with the Er compound.

Figure 3.5 summarizes the measurements of the SC gap or SC OP by PCs in the title compounds. In general, the SC gap or OP values are placed close to the BCS value $\Delta = 1.76 k_B T_c$, taking into account the multiband behavior in the R = Er compound and the vanishing of the SC gap in the Ho borocarbide at $T^* \approx 5.6$ K. For comparison, also the SC gaps of the nonmagnetic borocarbides R = Y and Lu are presented in Fig. 3.4. A two-gap state is established for R = Lu [15] and a strong anisotropy of the SC gap (probably due to multiband state) is observed for R = Y [24]. It is seen that the OPs in Er and Lu compounds have close values, but in the case of Er the larger OP contribution to dV/dI dominates at low temperature and then this contribution decreases (by a factor 5) with increasing temperature [14], while for R = Lu similar behavior is observed for the smaller OP [15].

3.4 Conclusions

The SC gap or SC OP was studied using normal metal-superconductor PCs for a series of magnetic rare-earth nickel borocarbide superconductors. For the first time, the existence of two SC OPs in the magnetic compound ErNi$_2$B$_2$C has been shown. Moreover, a distinct decrease of both the OPs is observed as the temperature is lowered below T_N. For the R = Ho and Dy compounds with commensurate AF order, the SC gap has a BCS like behavior in the AF state, while for R = Tm the gap starts to decrease by approaching a magnetic state with incommensurate AF order. Note that the Er compound has an incommensurate AF order and the OPs start to decrease with decreasing temperature slightly above T_N as well. Thus, the discrepancy in the magnetically ordered state between $R =$ Ho, Dy (commensurate state) and $R =$ Er, Tm (incommensurate state) results in a different SC gap or OP behavior. More extensive directional PC measurements for $R =$ Dy, Ho and Tm are desirable to check the presence of multigap superconductivity in these compounds as well.

Acknowledgments This work was supported by the National Academy of Sciences of Ukraine. The authors thank G. Behr, P. C. Canfield, K. Nenkov, and D. Souptel for the long-term collaboration in the field of rare-earth nickel borocarbide investigations and experimental assistance. Two of us, Yu. G. N. and O. E. K., thank the Alexander von Humboldt Foundation for support.

References

1. K.-H. Müller, M. Schneider, G. Fuchs, S.-L. Drechsler, in *Handbook on the Physics and Chemistry of Rare Earths*, vol. 38, Chap. 239, ed. by K.A. Gschneidner Jr., J.-C. Bünzli, V.K. Pecharsky (North-Holland, Amsterdam, 2008)
2. S.V. Shulga, S.-L. Drechsler, G. Fuchs, K.-H. Müller, K. Winzer, M. Heinecke, K. Krug, Phys. Rev. Lett. **80**, 1730 (1998)
3. Yu. G. Naidyuk, I.K. Yanson, *Point-Contact Spectroscopy*, vol. 145, Springer Series in Solid-State Sciences (Springer Science Berlin, 2005)
4. L.F. Rybaltchenko, I.K. Yanson, A.G.M. Jansen, P. Mandal, P. Wyder, C.V. Tomy, D.McK. Paul, Physica B **218**, 189 (1996)
5. I.K. Yanson, V.V. Fisun, A.G.M. Jansen, P. Wyder, P.C. Canfield, B.K. Cho, C.V. Tomy, D.McK. Paul, Fiz. Nizk. Temp. **23**, 951 (1997); Low Temp. Phys. **23**, 712 (1997)
6. L.F. Rybaltchenko, I.K. Yanson, A.G.M. Jansen, P. Mandal, P. Wyder, C.V. Tomy, D.McK. Paul, Europhys. Lett. **33**, 483 (1996)
7. L.F. Rybaltchenko, A.G.M. Jansen, P. Wyder L.V. Tjutrina, P.C. Canfield, C.V. Tomy, D.McK. Paul, Physica C **319**, 189 (1999)
8. I.K. Yanson, in *Symmetry and Pairing in Superconductors*, ed. by M. Ausloos S. Kruchinin (Kluwer, The Netherlands, 1999), pp. 271–285
9. I.K. Yanson, N.L. Bobrov, C.V. Tomy, D.McK. Paul, Physica C **334**, 33 (2000)
10. I.K. Yanson, in *Rare Earth Transition Metal Borocarbides (Nitrides): Superconducting, Magnetic and Normal State Properties*, vol. 14, ed. by K.H. Müller V. Narozhnyi (Kluwer, The Netherlands, 2001), pp. 95–108
11. Yu.G. Naidyuk, O.E. Kvitnitskaya, I.K. Yanson, G. Fuchs, K. Nenkov, A. Wälte, G. Behr, D. Souptel, S.-L. Drechsler, Physica C **460–462**, 105 (2007)
12. Yu.G. Naidyuk et al., Phys. Rev. B **76**, 014520 (2007)
13. Yu.G. Naidyuk, D.L. Bashlakov, N.L. Bobrov, V.N. Chernobay, O.E. Kvitnitskaya, I.K. Yanson, G. Behr, S.-L. Drechsler, G. Fuchs, D. Souptel, D.G. Naugle, K.D.D. Rathnayaka, J.H. Ross Jr, Physica C **460–462**, 107 (2007)
14. N.L. Bobrov, V.N. Chernobay, Yu.G. Naidyuk, L.V. Tyutrina, D.G. Naugle, K.D.D. Rathnayaka, S.L. Bud'ko, P.C. Canfield I.K. Yanson, Europhys. Lett. **83**, 37003 (2008)
15. N.L. Bobrov, S.I. Beloborod'ko, L.V. Tyutrina, I.K. Yanson, D.G. Naugle, K.D.D. Rathnayaka, Phys. Rev. B **71**, 014512 (2005); N.L. Bobrov, S.I. Beloborod'ko, L.V. Tyutrina, V.N. Chernobay, I.K. Yanson, D.G. Naugle, K.D.D. Rathnayaka, Fiz. Nizk. Temp. **32**, 641 (2006); Low Temp. Phys. **32**, 489 (2006)
16. G.E. Blonder, M. Tinkham, T.M. Klapwijk, Phys. Rev. B **25**, 4515 (1982)

17. S.I. Beloborodko, Fiz. Nizk. Temp. **29**, 868 (2003); Low Temp. Phys **29**, 650 (2003)
18. K.D.D. Rathnayaka, D.G. Naugle, B.K. Cho, P.C. Canfield, Phys. Rev. B **53**, 5688 (1996)
19. T. Watanabe, K. Kitazawa, T. Hasegawa, Z. Hossain, R. Nagarajan L.C. Gupta, J. Phys. Soc. Jpn. **69**, 2708 (2000)
20. T. Baba, T. Yokoya, S. Tsuda, T. Kiss, T. Shimojima, K. Ishizaka, H. Takeya, K. Hirata, T. Watanabe, M. Nohara, H. Takagi, N. Nakai, K. Machida, T. Togashi, S. Watanabe, X.Y. Wang, C.T. Chen, S. Shin, Phys. Rev. Lett. **100**, 017003 (2008)
21. K. Machida, K. Nokura T. Matsubara, Phys. Rev. B **22**, 2307 (1980)
22. H. Suderow, P. Martinez-Samper, N. Luchier, J.P. Brison, S. Vieira, P.C. Canfield, Phys. Rev. B **64**, 020503(R) (2001)
23. M. Crespo, H. Suderow, S. Vieira, S. Bud'ko, P.C. Canfield, Phys. Rev. Lett. **96**, 027003 (2006)
24. D.L. Bashlakov, Yu.G. Naidyuk, I.K. Yanson, G. Behr, S.-L. Drechsler, G. Fuchs, L. Schultz D. Souptel, J. Low Temp. Phys. **147**, 335 (2007)

Chapter 4
Peculiarities of Supershort Light Pulses Transmission by Thin Semiconductor Film in Exciton Range of Spectrum

P.I. Khadzhi, I.V. Beloussov, D.A. Markov, A.V. Corovai, and V.V. Vasiliev

Abstract Taking into account the exciton–photon and elastic exciton–exciton interactions we investigated peculiarities of transmission of supershort light pulses by thin semiconductor films. We predict the appearance of time dependent phase modulation and dynamical red and blue shifts of transmitted pulse.

4.1 Introduction

Studies of the unique optical properties of thin semiconductor films (TSF) induce a raised interest because of the great prospects of the practical applications. The nonlinear relation between the field of an electromagnetic wave propagating through a TSF and polarization of the medium gives rise to a number of interesting physical effects both under the stationary and nonstationary excitation. It is very important that the TSF has the property of the optical bistability in the transmitted and reflected light wave without any additional device. The peculiarities of the nonstationary interaction of the supershort pulses (SSP) of laser radiation with TSF was studied, which consists of two- and three-level atoms [1–7]. A number of new results were obtained in the investigation of the TSF transmission, taking into account the effects of exciton–photon interaction, the phenomenon of exciton dipole momentum saturation, the optical exciton–biexciton conversion, and the single-pulse and two-pulse two-photon excitation of biexciton from the ground state of the crystal [8–17]. The new possibilities of ultraspeed control by the transmission (reflection) of TSF were predicted, which can have great prospects of their utilization in the optical information processing systems. Therefore, further investigation into the peculiarities of the TSF transmission (reflection) in the exciton range of spectrum at a high level of excitation is of high priority.

4.2 Basic Equations

In this section, we present the main results of the theoretical investigation of nonlinear optical properties of the TSF in the conditions of the generation of high-density excitons by the supershort pulse of the resonant laser radiation with the amplitude of the electric field, E_i, and the frequency, ω, being in resonance with the exciton self-frequency, ω_0, normally incidents on the TSF with the thickness, L, which is much smaller than the light wave length, λ, but much more than the exciton radius, a_0. In the frame of these conditions, the elastic exciton–exciton interaction is the main mechanism of nonlinearity. We consider the process of pulse transmission through the TSF, taking into account the exciton–photon and elastic exciton–exciton interactions. We suppose that only one macrofilled mode of the coherent excitons and photons exists. We handle the problem using the Maxwell equations for the field and the Heisenberg (material) equation for the amplitude of the exciton wave of polarization. We proceed from the Keldysh equation [18] for the amplitude, a, of the exciton wave

$$i\hbar \frac{\partial a}{\partial t} = -\hbar (\Delta + i\gamma) a + g |a|^2 a - \frac{d_{\text{ex}}}{\sqrt{v_0}} E^+, \qquad (4.1)$$

where $\Delta = \omega - \omega_0$ is the resonance detuning, γ is the damping constant of the excitons, g is the constant of the elastic exciton–exciton interaction, d_{ex} is the exciton dipole momentum transition from the ground state of the crystal, v_0 is the volume of the unit cell of the crystal, and E^+ is the positive-frequency component of the electric field of the wave propagating through the TSF. The parameters d_{ex} and g contained in (4.1) are defined by the expressions

$$d_{\text{ex}} = (\varepsilon_b v_0 \hbar \omega_{\text{LT}}/4\pi)^{1/2}, \quad g = (26/3) \pi I_{\text{ex}} a_0^3, \qquad (4.2)$$

where ε_b is the background dielectric constant, a_0, ω_{LT}, and I_{ex} are the radius, longitudinal–transversal frequency, and the binding energy of the exciton respectively. The (4.1) for the bulk crystal corresponds to the mean field approximation, the applicability of which was considered in [18, 19]. Following [1, 3, 7, 8, 12, 13, 15, 17] the boundary conditions for the tangential components of the fields of the propagating pulses, we can obtain the electromagnetic relationship between the fields of incident, E_i, transmitted, E_t, and reflected, E_r, waves and the amplitude, a, of the exciton wave of polarization in the form

$$E_t^+ = E_i + i \frac{2\pi \omega L}{c} \frac{d_{\text{ex}}}{\sqrt{v_0}} a, \quad E_t^+ = E_i + E_r^+, \qquad (4.3)$$

where L is the film thickness and c is the light velocity in vacuum. We present the macroscopic amplitudes as the products of slowly varying in time envelopes and fast-oscillating exponential functions with the frequency ω. By introducing the normalized quantities

4 Peculiarities of Supershort Light Pulses Transmission by Thin Semiconductor Film

$$A = \sqrt{n_0}a, \quad F_t^+ = E_t^+/E_t^+ E_{\text{eff}}, \quad F_i = E_i/E_{\text{eff}}, \quad \tau = t/\tau_0,$$
$$\delta = \Delta\tau_0, \quad \Gamma = \tau_0\gamma, \tag{4.4}$$

we can rewrite the (4.1) and (4.3) in the form

$$i\frac{\partial A}{\partial \tau} = -(\delta + i\Gamma) A + |A|^2 A - iA - F_i, \tag{4.5}$$

$$F_t^+ = F_i + iA, \quad F_r^+ = iA, \tag{4.6}$$

where the characteristic time parameter τ_0 of the TSF response, the effective field strength, E_{eff}, and the exciton density, n_0, are defined by the expressions

$$\tau_0^{-1} = 2\pi\omega L d_{\text{ex}}^2/(c\hbar v_0), \quad n_0 = \hbar/(g\tau_0), \quad E_{\text{eff}} = \hbar\sqrt{n_0 v_0}/(d_{\text{ex}}\tau_0). \tag{4.7}$$

Let us estimate the values of parameters for CdS-like TSF with the thickness $L = 10^{-6}$ cm, using $\varepsilon_b = 9.3$, $v_0 = 1.25 \times 10^{-22}$ cm^{-3}, $a_0 = 2.8$ nm, $\hbar\omega_{\text{LT}} = 1.9$ meV, $I_{\text{ex}} = 29$ meV, and $\gamma = 10^{11}$ s^{-1} [20–24]. We have obtained $\Gamma = 5.7 \times 10^{-2}$, $d_{\text{ex}} = 0.53 \times 10^{-18}$ erg$^{1/2}$ cm$^{3/2}$, $g = 2.8 \times 10^{-32}$ erg cm^3, $\tau_0 = 5.7 \times 10^{-13}$ s, $n_0 = 6.3 \times 10^{16}$ cm^{-3}, and $E_{\text{eff}} = 2.8 \times 10^3$ V/cm, which corresponds to the intensity 10^3 W/cm^2.

Presenting the functions $A(\tau)$ and $F_t^+(\tau)$ as a sum of the real and imaginary parts, that is, $A = x + iy$, $F_t^+ = F + iG$ and supposing that the envelope of the incident pulse, F_i, is a real function of time, we can write the (4.5) and (4.6) in the form

$$\dot{x} = -(1 + \Gamma) x - (\delta - z) y, \tag{4.8}$$
$$\dot{y} = -(1 + \Gamma) y + (\delta - z) x + F_i, \tag{4.9}$$
$$F = F_i - y, \quad G = x, \tag{4.10}$$

where $z = x^2 + y^2$ is the normalized density of excitons, for which we can obtain the equation too

$$\dot{z} = -2(1 + \Gamma) z + 2F_i y. \tag{4.11}$$

We see that the amplitude of transmitted wave through TSF (as well as the amplitude of exciton polarization) is the phase-modulated function of time. In the linear approximation, the solution for the function $A(\tau)$ is the damped function of time

$$A = iF_0 (1 + \Gamma - i\delta)^{-1} \{1 - \exp[-(1 + \Gamma - i\delta) \tau]\}, \tag{4.12}$$

where the damping constant is equal to $1 + \Gamma$. In (12), the exponentially damping factor is conserved even though $\Gamma = 0$, i.e., if the exciton state does not decay. This is due to the fact that the film is the open system and the emergence of the radiation presents the additional "dissipative" mechanism in the system of excitons. In the limit $\Gamma \ll 1$ ($\gamma \ll \tau_0^{-1}$), the real dissipation due to the outflow of excitons from the coherent macrofilled mode to the incoherent one is not accomplished.

From (4.8)–(4.11), we obtain the dependence of the stationary density of excitons z_s on the field amplitude $F_i = F_0 = $ Const of the rectangular incident pulse. The dependence is defined by the solution of the cubic equation

$$z_s \left[(z_s - \delta)^2 + (1 + \Gamma)^2 \right] = F_0^2. \tag{4.13}$$

For the detuning $\delta < \delta_c = \sqrt{3}(1 + \Gamma)$ the density of excitons monotonously increases when the amplitude, F_0, of the incident pulse increases. The dependence $z_s(F_0)$ in this case is the nonlinear, but single-valued one. At high level of excitations, the dependence $z_s(F_0)$ is nonlinear, multivalued and is characterized by the hysteresis when $\delta > \delta_c$. For the values of the field amplitude, $F_\pm$, the jumps appear in the behavior of the function $z_s(F_0)$ at $z = z_\pm$, where

$$z_\pm = \frac{1}{3} \left[2\delta \pm \sqrt{\delta^2 - 3(1 + \Gamma)^2} \right], \tag{4.14}$$

$$F_\pm^2 = \frac{2}{27} \left[\delta \left(\delta^2 + 9(1 + \Gamma)^2 \right) \pm \left(\delta^2 - 3(1 + \Gamma)^2 \right)^{3/2} \right]. \tag{4.15}$$

For the critical resonance detuning $\delta = \delta_c = \sqrt{3}(1 + \Gamma)$, we obtain $z_c = 2(1 + \Gamma)/\sqrt{3}$ and $F_c = z_c^{3/2}$. The cyclic change of the amplitude, F_i, of incident pulse results in the jump-like change of the exciton density in film forming the hysteretic behavior for $F_+ < F_0 < F_-$. The shift of the exciton level for the increasing excitation intensity is the main physical reason for the hysteretic dependence $z(F_0)$. An investigation of these solutions concerning the stability upon the small deviation from the stationary values leads to a conclusion that the instability of the stationary state for z_s takes place in the range $z_- < z < z_+$, which corresponds to the middle part of the hysteretic curve $z_s(F_0)$. The system of (4.8)–(4.11) has three stationary solutions, two of which are the specific points such as stable node (for $\delta/3 < z < \delta$) or stable focus (for $z < \delta/3$ and $z > \delta$) and the third point is a saddle. The phase portrait of the dynamic system (4.8) and (4.9) in the region of the trivaluedness is presented on Fig. 4.1.

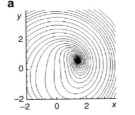
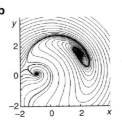

Fig. 4.1 The phase portrait of the dynamical system under the investigation for the detunings $\delta = 0$ (**a**) and $\delta = 5$ (**b**)

4.3 Discussion of Results of Numerical Solutions

We now discuss the peculiarities of the nonstationary behavior of the system supposing that at the initial moment of time ($\tau = 0$) the crystal was in the ground state, that is, $x|_{\tau=0} = y|_{\tau=0} = z|_{\tau=0} = 0$. The behavior of the system is defined by the level of excitation and detuning. First, let us consider the case of exact resonance: $\delta = 0$. Let the rectangular pulse with the amplitude $F_i = F_0 =$ Const incidents on the TSF. In Fig. 4.2, we presented the time evolution of the exciton density in the TSF and the amplitude of transmitted pulse, $|F_t|$, for the different values of the field amplitude, F_0, of the incident pulse. At low levels of excitation, the exciton density monotonously increases in time: $z = F_0^2/(1+\Gamma)^2 \times \left(1 - e^{-(1+\Gamma)\tau}\right)^2$, and at long times ($\tau \gg 1$), it reaches the stationary value $z_s \approx F_0^2$. Exciton density at the initial stage of evolution accelerates when the level of excitation increases, and the small-amplitude oscillations of the density with a long period appear. The further increase of the amplitude, F_0, of the incident pulse leads to a sharp increase of the exciton density at the initial stage of the evolution and to the pronounced oscillations of the exciton density. The amplitude of oscillations increases with F_0 and monotonously decreases in time for fixed F_0. As for the period of oscillations of exciton density, it monotonously decreases when the level of excitation increases. Then the oscillations of exciton density gradually decay and the steady state is reached with the exciton density z_s (Fig. 4.2a).

The obtained peculiarities of the behavior of the function $z(t)$ determine the peculiarities of the time evolution of the field amplitude, $|F_t|$, of transmitted radiation. At low levels of excitation, the field amplitude, $|F_t|$, monotonously decreases in time and approaches the stationary transmission. When $\Gamma = 0$, the TSF is closed in transmission and all incident radiation is reflected. In this limit, the TSF can play the role of ideal mirror for SSP with the duration $\tau_{\text{pulse}} \ll \Gamma^{-1}$. When the level of excitation increases, the week pronounced oscillations with a long period appear after initial decrease of the function $|F_t(\tau)|$. The further increase of the incident amplitude, F_0, leads to the increase of the rate of change of the transmitted pulse at the initial evolution stage and to the appearance of a well-pronounced oscillations of the film transmittance. In this case, the period of oscillations decreases and the amplitude of oscillations increases with the increase of F_0 and even can happen that $|F_t| > F_0$ in the peak of the first maximum (Fig. 4.2d). It is due to the fact that the field of transmitted pulse is the sum of the field of incident pulse and the secondary field, which is generated by the exciton polarization. If both fields are changed in phase, then the amplitude of transmitted pulse appears to be more than the incident amplitude. For a fixed F_0, the amplitude of oscillations monotonously decreases in time and the steady state transmission with the amplitude $|F_t| \approx F_0$ is set.

Let us discuss now the peculiarities of the time evolution of the system for the detunings $\delta > \delta_c$ (Fig. 4.2b,e). In this case, at low levels of excitation, the exciton density very slowly increases depending on both F_0 and the time. At the initial stage of the time evolution, we observe a small increase of the exciton density. The same takes place in transmission too: the amplitude of transmitted pulse at first decreases

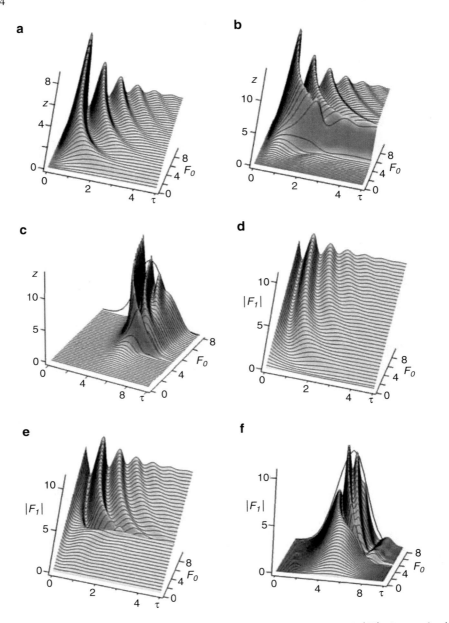

Fig. 4.2 The time evolution of the exciton density z (**a**, **b**, **c**) and amplitude $|F_t|$ of transmitted pulse (**d**, **e**, **f**) for the different values of the amplitude F_0 of incident rectangular (**a**, **b**, **d**, **e**) and Gauss (**c**, **f**) pulses and for the resonance detuning $\delta = 0$ (**a**, **c**, **d**, **f**) and $\delta = 5$ (**b**, **e**)

in time, one or two slightly appreciable oscillations appear and then the system transits into the steady state. However, this takes place up to the moment when the amplitude of incident pulse, F_0, will be equal to F_-, which corresponds to the jump from one branch to another of the hysteretic dependence of exciton density on F_0 in the stationary state (Fig. 4.2b,e). Immediately after the bifurcation, the time evolution of the system at the initial stage exhibits a very fast increase of the exciton density, which is then transformed into a pronounced oscillatory regime of evolution with the great amplitudes and small periods of oscillations. The amplitude of oscillations decreases in time and the system gradually reaches its steady state. The same phenomena take place in the transmission too: immediately after bifurcation, the regime with pronounced oscillations of the transmitted pulse appears. From Fig. 4.2b,e, we can see that for $\delta > \delta_c$, a transmittance with the amplitude of the first peak much more than the amplitude of incident pulse is possible.

We point out that the change of the amplitude of incident rectangular pulse and the resonance detuning leads to the generation of the complex forms of the transmitted pulses.

Let us discuss now the peculiarities of the nonstationary transmission of thin film for the case of normal incidence of Gauss-like pulse $F_i(\tau) = F_0 \exp(-t^2/T^2)$, where F_0 and T are the amplitude and half-width of pulse. In Fig. 4.2c,f, we presented the time evolution of the exciton density, z, and the amplitude, $|F_t|$, of transmitted pulse for different values of the amplitude, F_0, of incident pulse. We can see that the shape of the transmitted pulse is changed sufficiently with the change of amplitude, F_0. The small peak of the exciton density and transmitted pulse appear at small values of F_0. The maximum of transmitted pulse falls at the early stage with respect to the maximum of the incident pulse. When the value of F_0 increases the oscillations of the peak of exciton density appears, which synchronously changes the behavior of the transmitted pulse. The peak of this pulse gradually becomes narrow and it shifts to the front tail of the incident pulse, whereas at moments that correspond to the center or to the rear tail of the incident pulse, the new peaks appear one after another, the amplitude and the quantity of which increase with the increase of F_0. Far from the rear tail the weak pulse appears, which is due to the emission of a secondary radiation. The half-widths of new peaks are significantly small in comparison with the half-width of the incident pulse.

In Fig. 4.3, we presented the time evolution of the amplitude of transmitted pulse for the Gauss-like incident pulse and the transmission function, that is, dependence of the amplitude of transmitted pulse on the amplitude of incident one. We can point the difference between the stationary and nonstationary transmittances. The incident pulses with a large amplitude lead to the strong oscillations of the transmitted pulse, which represents the significant deviation from the steady state transmittance. Therefore we can affirm that the steady state transmittance is obtained using the values of the amplitudes of transmitted pulses, which are set after the transient stage, when all oscillations were terminated. The result of the experimental investigations of the transmission function using the supershort Gauss pulses will significantly differ from the theoretical results, which were obtained in the assumption of the

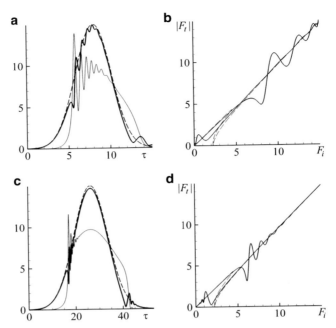

Fig. 4.3 Time evolution of the exciton density (*thin curve*) and amplitude of transmitted pulse $|F_t|$ (*thick curve*) (**a**, **b**) and the transmittance function F_t (F_0) in comparison with the steady-state transmittance for the Gauss pulses with half-widths T, equal 3 (**a**, **b**) and 10 (**c**, **d**)

stationary transmission. The multivaluedness of the transmittance will be due to the transient effects rather than due to the bistability in the steady-state regime.

As we pointed out, we have not succeeded in obtaining the nonlinear solution of (4.8)–(4.11). However, for one unique case we have obtained the solution to the exciton density, $z(\tau)$, under the action of the incident pulse with the shape $F_i = F_0 \exp(-3\tau)$, which has the form

$$z = \frac{F_0^{2/3}\left(\sqrt{3}-1\right)}{2^{2/3}} e^{-2\tau} \left[1 - \operatorname{cn}\left(\sqrt[4]{3}\left(\frac{F_0}{2}\right)^{2/3}\left(1-e^{-2\tau}\right)\right)\right] \quad (4.16)$$

$$\cdot \left[1 + \left(2-\sqrt{3}\right)\operatorname{cn}\left(\sqrt[4]{3}\left(\frac{F_0}{2}\right)^{2/3}\left(1-e^{-2\tau}\right)\right)\right]^{-1},$$

where $\operatorname{cn}(x)$ is the elliptic Jacoby function [25] with the modulus $k = \sqrt{2-\sqrt{3}}/2 = \sin(\pi/12)$. In Fig. 4.4, we presented the exciton density, z, and the amplitude, F_t, of transmitted pulse. We can see the appearance of the weak oscillations of these functions for the case of a strong pump.

Fig. 4.4 Time evolution of the exciton density z and amplitude F_t (*thin curve*) of transmitted pulse (*thick curve*) for the incident pulse of the form $F_i = F_0 \exp(-3\tau)$

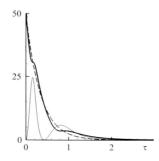

4.4 Conclusions

In conclusion, we studied the transmittance of TSF under the action the supershort pulses of resonant laser radiation, taking into account the exciton–photon and the elastic exciton–exciton interaction. We pointed out the differences between stationary and nonstationary transmissions and the possible experimental difficulties of their observation.

Acknowledgments This work was supported by the joint grants of the Academy of Sciences of Moldova and Russian Fund of Fundamental Investigations 06.03 CRF/06–02–90861 Mol_a, No. 08.820.05.039 RF and by the Project nr. 076/P from February 14, 2008 in the frame of State Program of Moldova.

References

1. V.I. Rupasov, V.I. Yudson, Kvant. Elektron. (Moscow) **9**, 2179 (1982)
2. V.I. Rupasov, V.I. Yudson, Zh. Eksp. Teor. Fiz. **93**, 494 (1987)
3. S.M. Zakharov, E.A. Manykin, Poverknost **2**, 137 (1988)
4. S.M. Zakharov, E.A. Manykin, Poverknost **7**, 68 (1989)
5. S.M. Zakharov, E.A. Manykin, Zh. Eksp. Teor. Fiz. **95**, 800 (1989)
6. S.M. Zakharov, E.A. Manykin, Zh. Eksp. Teor. Fiz. **95**, 1053 (1994)
7. A.M. Samson, Yu. A. Logvin, S.I. Turovets, Kvant. Electron.(Moscow) **17**, 1223 (1990)
8. P.I. Khadzhi, S.L. Gaivan, Zh. Eksp. Teor. Fiz **108**, 1831 (1995)
9. P.I. Khadzhi, S.L. Gaivan, Kvant. Electron.(Moscow) **22**, 929 (1995)
10. P.I. Khadzhi, S.L. Gaivan, Kvant. Electron.(Moscow) **23**, 837 (1996)
11. P.I. Khadzhi, S.L. Gaivan, Kvant. Electron.(Moscow) **24**, 532 (1997)
12. P.I. Khadzhi, L.V. Fedorov, Zh.Tekhn. Fiz **70**, 65 (2000)
13. P.I. Khadzhi, A.V. Corovai, Quant. Electron. **32**, 711 (2002)
14. P.I. Khadzhi, A.V. Corovai, Adv. Mater. **5**, 37 (2003)
15. P.I. Khadzhi, A.V. Corovai, D.A. Markov, Gauges Syst. **12**, 47 (2004)
16. P.I. Khadzhi, A.V. Corovai, D.A. Markov, Mold. J. Phys. Sci. **4**, 408 (2005)
17. P.I. Khadzhi, A.V. Corovai, D.A. Markov, V.A. Lichman, Proc. of SPIE, ICONO, **6259**, 62590M.1–9 (2006)
18. L.V. Keldysh, in *Problems of Theoretical Physics* (M. Nauka, Moscow, 1972), p. 433
19. Yu. I. Balkarei, A.S. Kogan, Pis'ma v JETP **57**, 277 (1993)
20. S.A. Moskalenko, D.W. Snoke *Bose–Einstein Condensation of Excitons and Biexcitons and Coherent Nonlinear Optics with Excitons* (Cambridge Univ. Press., 2000)

21. J. Pancov, *Optical Processes in Semiconductors* (Mir, Moscow, 1973)
22. V.G. Litovchenko et al., Fiz. Tekhn. Polupr. **36**, 447 (2002)
23. E.I. Rashba, M.D. Sturge, eds. *Excitons* (Nauka, Moscow, 1985)
24. M.S. Brodin, E.N. Miasnicov, S.V. Marisova, *Polaritons in Crystallooptics* (Kiev, Naucova Dumka, 1984)
25. I.S. Gradshteyn, I.M. Ryzhik, *Table of Integrals, Sums, Series, and Products* (Fizmatgiz, Moscow, 1963)

Part II
Nanomaterials and Nanoparticles

Chapter 5
Nanostructuring and Dissolution of Cementite in Pearlitic Steels During Severe Plastic Deformation

Y. Ivanisenko, X. Sauvage, I. MacLaren, and H.-J. Fecht

Abstract Strain-induced cementite dissolution is a well-documented phenomenon, which occurs during the cold plastic deformation of pearlitic steels. Recently new results that can shed additional light on the mechanisms of this process were obtained thanks to atom probe tomography investigations of pearlitic steel deformed by highpressure torsion (HPT). It was shown that the process of cementite decomposition starts by carbon depletion from the carbides due to defect motion; once enough carbon is robbed from the carbide it is thermodynamically destabilized resulting in rapid break-up. Additionally, it was shown that the carbon atoms do not really dissolve in the ferrite but that they segregate to the dislocations and grain boundaries of nanocrystalline ferrite.

5.1 Introduction

Pearlite, a lamellar structure comprising alternating platelets of ferrite and cementite (an iron carbide of formula Fe_3C), is the most important constituent of carbon steels, and provides their high strength. The interlamellar spacing in pearlite often varies from 100 to 500 nm, and cementite lamella thickness can be as small as 10 nm Hence fine pearlite can be regarded as a natural one-dimensional nanomaterial, which has been used by mankind for hundreds of years prior to the dawn of the "nanomaterials era" at the end of twentieth century. Cementite is a hard and brittle, covalently bonded compound consisting of three Fe atoms and one C atom assembled in a complex orthorhombic structure. It is a metastable compound under ambient conditions, and decomposes to a stable graphite and iron during extended annealing at high temperatures (many tens of hours). Despite this, it is very stable at the room temperature, and thus the steel metallurgy community was very doubtful about the early reports of its decomposition (or dissolution) as a consequence of cold plastic deformation. In fact, only recent measurements of atomic distributions in deformed pearlitic steel made using atom probe tomography finally confirmed the occurrence of strain-induced cementite dissolution.

Strain-induced dissolution of iron carbide cementite in pearlitic steels was first reported more than 20 years ago [1], and since then there have been a growing interest in this phenomenon. This has occurred for two main reasons. Firstly the issues related to the phase stability of pearlitic steels subjected to external forces are fundamental scientific interest. Secondly, however, it is also of great practical relevance in view of the huge practical importance of steels in construction and engineering. Steels are often subject to high strains either during fabrication (e.g. wire drawing or cold rolling) or in application. For example, rails often undergo very high strains in use due to high contact pressures at the wheel–rail contact surfaces. In fact, the so-called white etching layer (WEL) formed on the surface of rail tracks, which, because of its brittleness, leads to rapid erosion in use, is a nanostructured Fe–C alloy produced by the dissolution of cementite under the cyclic heavy plastic deformation at the rail–wheel contact [2, 3]. Cementite dissolution is always accompanied by a strong refinement of the ferrite grain size down to nanoscale, and the microhardness of this newly formed Fe–C alloy increases to values higher than that of conventional martensite [4, 5]. Moreover, its thermal stability is also better than that of martensite [5]. Thus, it could be imagined that the use of strain-induced cementite dissolution can offer perspectives to produce new structural states in this well-studied material yielding different properties to those normally available for steels.

Despite the abundance of experimental studies of strain-induced cementite dissolution and the associated nanostructuring, conducted on various pearlitic steels deformed in different conditions (including cold rolling [1, 6], wire drawing [7–9] ball milling [10–12], shot peening [13, 14], highpressure torsion (HPT) [4, 15, 16], friction and wear conditions in the contact area of the railway tracks and wheels [2, 3]), many fundamental questions about the atomistic mechanisms of dissolution and about the resulting distribution of the carbon atoms in the ferrite matrix are still not fully understood. Direct observation of the microstructure of carbides and the detection of very small concentrations of carbon is a very complicated task. Carbon is a light element and its content in pearlitic steels is usually less than 4 at.%; such low concentrations are difficult to detect and quantify with analytical transmission electron microscopy (TEM) using techniques such as energy dispersive X-ray analysis (EDX) or electron energy loss spectroscopy (EELS). Only thanks to the recent progress in atom probe tomography (APT), has it been possible to watch the fate of the carbon atoms as a consequence of the deformation. This combined with studies of the nanostructure and crystallographic structures using High-resolution TEM (HRTEM) has made possible the progress in our understanding that will be discussed in this paper.

5.2 Experimental

The majority of the results discussed in this paper were obtained on a pearlitic steel with composition: Fe–0.8 wt.% C–1.2 wt.% Mn subjected to HPT. The HPT process was carried out under a pressure of 7 GPa. In this procedure, the sample was placed between an upper immobile and lower rotatable Bridgman anvils (Fig. 5.1). The

5 Nanostructuring and Dissolution of Cementite in Pearlitic Steels

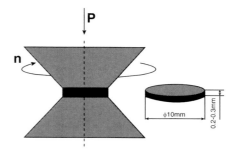

Fig. 5.1 Illustration of the HPT setup and of the processed specimen

shear was realized by turning one anvil relative the other at a speed $\omega = 1$ rpm. As with conventional torsion deformation, the shear strain, γ, can be estimated by means of the equation:

$$\gamma = \frac{2\pi N R}{h} \qquad (5.1)$$

where R is the distance from the sample centre, N is the number of the anvil rotations, and h is the thickness of the sample.

The samples twisted for different number of turns were then characterized using optical, scanning and TEM, and APT.

APT is based on the field-evaporation of atoms. This is a projection microscope combined with a time of flight mass spectrometer [17]. Huge electric fields are required to field-evaporate atoms (several volts per angstrom). However, this is easily achieved by applying a voltage of a few kilovolts to a specimen prepared as a sharp needle with a radius at the apex in the range of 10–50 nm. Due to the high electric field, atoms from the specimen surface are ionized and then radially accelerated by the electric field. They are collected on a position-sensitive detector located in front of the sample. The evaporation is performed in a UHV chamber at cryogenic temperatures (in a range of 20–100 K) and it is controlled by short electric pulses (duration of about one nanosecond, frequency in a range of 1–100 kHz), so that the time of flight could be accurately measured and the mass over charge ratio of the ions precisely estimated. The mass resolution depends on the flight length, but is usually improved by energy deficit compensation systems like reflectrons [18]. This way, isotopes of a given element can be separated, and for the best detectors with a low background noise [19], the detection limit can be as low as 20 at. ppm. Specimens are evaporated atomic layer after atomic layer, and thus, APT is a destructive technique that is not sensitive to contamination. Evaporated ions follow a quasi-stereographic projection that can be determined thanks to field ion microscopy (FIM) [17]. Impacts on the detector are precisely measured and then the original position in the evaporated material is computed thanks to a reverse projection. The spatial resolution of the instrument is 0.3 nm, and in some specific cases can be as good as 0.1 nm, making possible the reconstruction of atomic planes in crystalline metals. APT samples could be prepared by standard electropolishing techniques [17] or by ion milling using focused ion beams [20].

Thus, the APT is the only analytical microscope providing the three-dimensional distribution of chemical species at the atomic scale. During the last twenty years, it has been widely used in various metallic alloys to measure the composition of nanoscale particles and to display their 3D distribution. It is also a powerful instrument to reveal chemical gradients or to highlight segregation along structural defects like grain boundaries or dislocations [21].

This technique is perfectly adapted for the investigation of the decomposition of the Fe_3C phase in severely deformed pearlitic steels. Indeed, to understand the physical mechanisms of this phase transformation, the following are needed: (1) the evolution of the shape and the size of the carbides (Fe_3C lamellae); (2) the possible change of the carbon concentration in the carbides; (3) the carbon gradient at the Fe_3C/ferrite interfaces; and (4) the distribution of the carbon atoms within the nanostructure (i.e., homogeneous solid solution or segregation along defects). One should note that contrary to EELS, the main advantage of APT is that carbon concentration measurements are not affected by contamination.

5.3 Results and Discussion

5.3.1 Changes in the Microstructure and in Phase Composition of the Pearlitic Steel During HPT

The evolution of the pearlite morphology during HPT deformation in a wide range of shear strains $60 < \gamma < 600$ and its relationship to the cementite decomposition have been studied in [4]. The initial microstructure was fully pearlitic with randomly oriented pearlite colonies with a thickness of cementite lamellae of 40 nm and an interlamellar spacing of 250 nm (Fig. 5.2a). In a wide range of shear strains between 60 and 200, gradual rotation of pearlite colonies in the shear direction accompanied by thinning and elongation of cementite platelets was observed as shown in Fig. 5.2b, c.

After a certain shear strain, nonetching areas started to appear in colonies oriented parallel to the shear direction (Fig. 5.2c). It is known that carbon usually increases the etching resistance of iron; iron martensite does not etch at all, for example. Thus, the conclusion was made that in these nonetching areas, carbides have been mainly dissolved. The formation of nonetchable areas starts from the periphery of the HPT specimen, i.e., in areas with the highest shear strain, (5.1) and gradually progresses with each turn toward the center of the sample.

The microstructure of ferrite changes from single crystal in nondeformed lamellae (Fig. 5.3a) to cellular for intermediate stages of HPT deformation (Fig. 5.3b) and finally it transforms to a homogenous nanocrystalline one with a mean grain size of ferrite of 10–20 nm (Fig. 5.3c) [22]. This grain size is one order of magnitude smaller than that achieved in pure iron deformed in similar conditions [23]. Further straining does not lead to notable changes in the microstructure and the

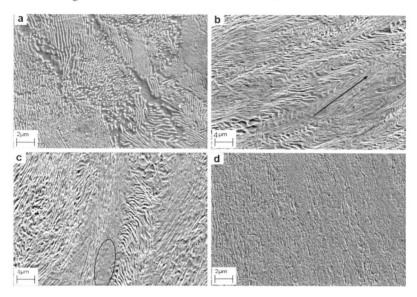

Fig. 5.2 Typical evolution of the cementite morphology in the course of HPT: (**a**) initial state – cementite colonies are randomly oriented; (**b**) after one rotation the majority of the colonies is aligned parallel to the shear direction (shown with *arrow*) (**c**) after two rotations areas, resistant to etching appear (shown by an *oval*); (**d**) after five rotations no cementite lamellae can be resolved in the microstructure. All images were recorded using secondary electron contrast in scanning electron microscopy on specimens etched with Nital

microhardness of samples shows a tendency to stabilize. Furthermore, a careful analysis of selected area diffraction (SAD) patterns had revealed a few weaker reflections due to cementite (marked **C) (Fig. 5.3d), and dark-field imaging of cementite has showed occasional very fine equiaxed cementite particles distributed uniformly in the structure [22].

The thinning and elongation of cementite lamellae in direction of deformation accompanying by partial cementite decomposition has been observed also for cold rolling [1, 6] and wire drawing [7–9]. However, for these processes, the nanocrystalline steady stage with almost complete cementite dissolution has never been achieved. On the other hand, such treatments as ball milling [10–12], shot peening [13, 14], and wear processes on surfaces of railway tracks and wheels [2, 3] usually result in nanostructuring of ferrite and complete Fe_3C dissolution.

It is clear from these and other studies that the evolution from cellular dislocation structures in the ferrite to the equiaxed nanostructure is correlated with the break-up of the cementite and partial dissolution of the carbon, the details of the process were unclear until recently. Evidence collected over the last few years has now made it possible to better understand these processes. At relatively low levels of deformation, it is already possible to break up the cementite lamellae, especially for thicker cementite lamellae, and this is probably caused by the passage of dislocations along a particular slip plane. One example of such broken cementite lamellae is shown in

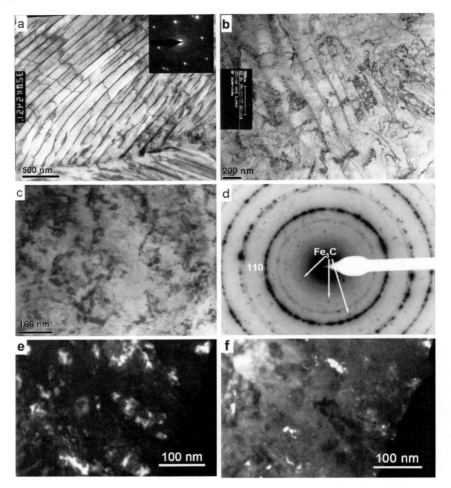

Fig. 5.3 TEM images of the micro/nano-structure of pearlitic steel: (**a**) in initial state, bright field image; and after HPT deformation: (**b**) for three rotations, bright field image showing formation of structure of misoriented cells in ferrite; (**c**) after five rotations, bright field image; (**d**) selected area diffraction pattern after five rotations showing some cementite reflections are labeled

Fig. 5.4a, this being from a rail steel in a lightly deformed region well away from the contact surface where severe deformation is experienced. On further deformation, for instance after one rotation in HPT at a radius of about 3 mm (shear strain $\gamma \approx 60$), fine cementite lamellae are frequently bent at a micron scale (Fig. 5.4b) at the same time as extensive dislocation cell formation seems to be happening in the ferrite. This seems to correlate with the breakup of the cementite lamellae into nanocrystals (Fig. 5.4c). Early studies by HRTEM seem to show the intermingling of cementite and ferrite in the "cementite" lamellae at this stage [24] and it would seem likely that dislocation processes are contributing to transport of material into and out

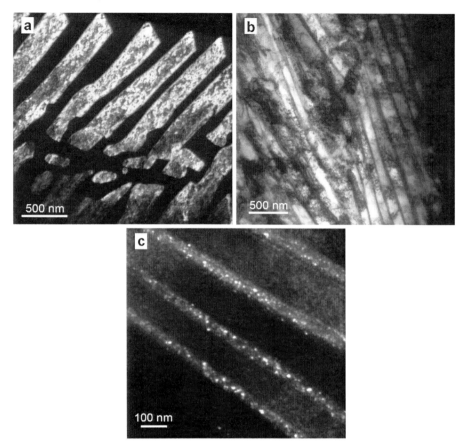

Fig. 5.4 Nanostructural processes involved in the breakup of cementite lamellae illustrated using dark- and bright-field TEM: (**a**) Breakup of lamellae, probably due to the passage of a shear band in the ferrite; (**b**) bending of lamellae on the micron scale by high pressure torsion; (**c**) breakup of lamellae into nanocrystals

of the lamellae – a process that will both break up the lamellae into nanocrystals and ultimately destroy the lamellae completely.

5.3.2 Variations of the Chemical Composition of Carbides

The conclusions about the gradual dissolution of cementite during HPT were fully confirmed by APT investigations and important details of this process have been revealed [25, 26]. It appears that decomposition of cementite at plastic deformation occurs via formation of a continuous range of nonstoichiometric cementite-like

Fig. 5.5 3D reconstruction of an analyzed volume in the pearlitic steel processed one turn by HPT (**a**) Only carbon atoms are plotted to exhibit a cementite lamellae. The carbon concentration profile was computed with a two nanometers sampling box across the ferrite/cementite interface (**b**) Reprinted from [26]

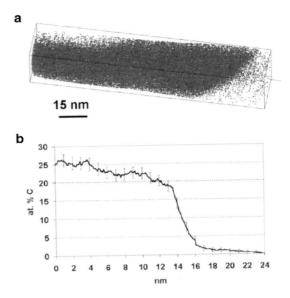

phases with C concentration decreasing from the center of the lamella to the interphase boundary. The larger the applied strain, the stronger the deviation of carbon content from stoichiometric. Figure 5.5a shows a region in a sample deformed for shear strain $\gamma = 60$, and only carbon atoms are shown in the Figure.

This is most probably a small part of a former cementite lamella that has been fragmented and plastically deformed. The composition profile computed across the α-Fe–cementite interface exhibits a sharp carbon gradient at this interface (Fig. 5.4b). In the far left of the profile, the carbon concentration is 25 at.%, as expected for cementite. However, along the interface there is a 10 nm thick layer with a carbon content in a range of 20–25 at.%, which could be attributed to off-stoichiometry cementite. On the ferrite side, a large carbon gradient appears: there is about 2 at.% carbon in the ferrite close to the interface and this value slowly decreases down to zero at a distance of about 8 nm from the interface.

APT investigations of pearlitic steel samples HPT-deformed to higher strains have confirmed the revealed tendency to formation of nonstoichiometric cementite as a result of plastic deformation. Figure 5.6a represents a region in the sample deformed to shear strain $\gamma = 300$ with lamellar-like structure with an interlamellar spacing in a range of 10–20 nm. The composition profile computed across these carbon-rich lamellae clearly shows that they contain a significant amount of carbon, about 4 at.% (Fig. 5.6b). However, only few regions with this kind of lamellar structure were revealed (less than 20 vol.%). Other colonies have been completely transformed into an equiaxed structure made of nano-scaled α-Fe grains stabilized by carbon atoms segregated along grain boundaries (Fig. 5.6c, d). The distribution of carbon atoms in this nanostructure is shown in Fig. 5.6c and revealed by the composition profile computed through this volume (Fig. 5.6d). A nanoscale cementite

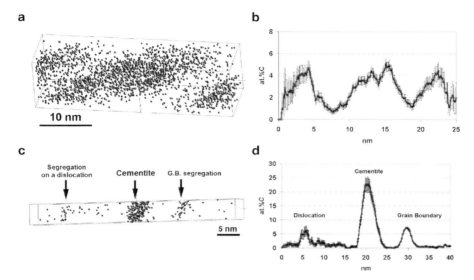

Fig. 5.6 (**a, c**) 3D reconstruction of two analyzed volumes ($6\times 6\times 43$ nm^3 and $11\times 11\times 34$ nm^3) in the pearlite processed by HPT ($\gamma = 300$). Only carbon atoms are plotted to show three carbon rich lamellae; (**b**) carbon concentration profiles computed from the left to the right across carbon rich lamellae exhibited in Fig. 5.4a. The carbon concentration of these lamellae is only about 4 at.%; (**d**) carbon concentration profile computed across the 3D volume in Fig. 5.7c [25]

Table 5.1 Comparison between the wire drawing process and HPT

	Strain rate	Temperature	Deformation process	Shape	Deformation mode	Hydrostatic pressure	Typical strain
Wire drawing	+ + + + + +		Discontinuous	Change	Multiple shear	low	3–4
HPT	− − − − − −		Continuous	Does not change	Pure shear	high	200–500

particle is located in the middle of the volume. As shown by the profile, the cementite is slightly off-stoichiometry and contains between 20 and 25 at.% C. On the left and right of the particle, segregation of carbon atoms along two planar defects is clearly exhibited.

Very similar data about the transformation of the Fe$_3$C lamellae in drawn pearlitic steels have been reported [9, 27]. The decomposition of Fe$_3$C in severely deformed pearlite was revealed for the first time by Mössbauer spectroscopy in cold-drawn pearlitic steels [1]. During the last ten years, these materials have been widely investigated by APT and although wire drawing and HPT are two very different processes, there are numerous similarities with the features reported in the present study.

The main differences between HPT and wire drawing are listed in Table 5.1. The strain level reached by HPT is usually much higher, but the strain rate during drawing is higher. This higher strain rate (with drawing speeds up to 500 m/min) may induce some significant temperature increase even in wet drawing operations, while this is very unlikely to occur during HPT. During HPT, the plastic deformation is continuously applied to the sample, while during drawing the wire diameter is reduced step by step through a full set of dies. The hydrostatic pressure is much higher during HPT (to avoid slip between the anvils and the sample), but the deformation mode is simpler (pure shear).

Cold-drawn pearlitic steel wires are among the strongest commercial steels, and they typically exhibit a yield stress higher than 3 GPa. Such a high strength is usually attributed to their unique nanostructure: pearlite colonies are strongly elongated along the wire axis during the drawing process and the interlamellar spacing is reduced to about 20 nm. During HPT, a similar phenomenon occurs: although the Fe_3C is supposed to be brittle since it does not have five independents slips systems, the lamellae deform plastically to some extent (although they do not remain as single crystals – see Fig. 5.4). Depending on their original orientation, they eventually bend and then they are strongly elongated along the shear direction. For the lowest strain, just as in drawn pearlite, the resulting material is a nanoscale lamellar structure with an interlamellar spacing in a range of 10–50 nm [4].

The strain-induced decomposition of the carbides in drawn pearlite typically starts for a true strain range between 2 and 3, and some authors have reported that carbides are fully decomposed at a strain of about 5.5 [8]. This phenomenon has been widely investigated by APT since the late 1990s [8, 9, 28–31]. However, the driving force and the kinetics of this phase transformation are still controversial. The main difficulty for the interpretation of experimental data is related to the random orientation of pearlite colonies prior to the drawing process. Depending on the orientation of lamellae with the drawing axis, the deformation of both α-Fe and Fe_3C is more or less pronounced, which gives rise to a wide range of interlamellar spacings in the drawn wire and to the experimentally observed heterogeneous decomposition of cementite.

Above a critical strain, the Fe_3C lamellae become so thin (only a few nanometres) that they cannot sustain more plastic deformation and fragmentation occurs together with decomposition (Fig. 5.7), in agreement with the TEM results of Fig. 5.4c. However, it is worth noticing that the lamellar nanostructure aligned along the wire axis is retained. In HPT samples, the situation is very different, especially because the strain level is much higher and the strain rate is much smaller. This gives rise to more opportunities for carbon atoms to interact with dislocations, and also more time to diffuse. Thus a full redistribution of carbon atoms happens and the lamellar structure may disappear completely.

In conclusion, APT analyses confirm that cementite is decomposed by severe plastic deformation, and the data provide new information about the decomposition mechanisms. Moreover, APT data clearly confirmed the TEM observation that showed that even after shear deformation $\gamma = 300$ cementite decomposition remains incomplete (Fig. 5.7). However, the remaining Fe_3C particles are very small and noticeably substoichiometric.

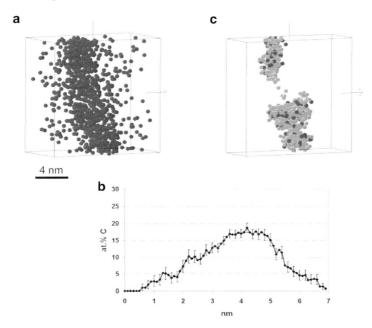

Fig. 5.7 (a) 3D reconstruction of a volume $(8 \times 8 \times 4\,\text{nm}^3)$ analyzed by APT in a cold drawn pearlitic steel (true strain 2.3), only C atoms are displayed to exhibit a Fe_3C lamella with a thickness of only 3 nm. (b) Composition profile computed across the carbon rich lamella of the showing that the average carbon content is lower than 25 at.% (sampling box: 1 nm). (c) Same volume as (a), but the data were filtered to exhibit only the regions containing more than 20 at.% C, i.e., some near-stoichiometric Fe_3C clusters within the carbon rich lamella (C atoms in black and Fe atoms in grey)

5.3.3 Distribution of Released Carbon Atoms in the Microstructure

Since the discovery of the phenomenon of deformation induced cementite decomposition, the question about the distribution of released C atoms was a real puzzle. The equilibrium solubility of C in α-Fe is very low at the room temperature, and dissolution of even small amounts of C should lead either to notable increase of lattice parameter or to the formation of tetragonal martensite. However, neither shifts nor the splitting of XRD peaks has been ever observed. It was proposed that C atoms are located in places where they will not cause lattice expansion and will not lead to a change of lattice parameter, i.e., the sites of decreased atomic density as the core of the dislocation or free volume of a high-angle grain boundary [4]. The results recently obtained using 3D AP and presented earlier have confirmed this suggestion, clearly showing that the released carbon atoms segregate to dislocations, grain and cell boundaries in ferrite (Fig. 5.6c, d) [26]. Similar distributions of carbon along the boundaries of nanocrystalline grains were found in ball-milled pearlitic steel after strain-induced cementite dissolution [32]. In some cases, the combination of high strains and the liberated carbon content held at such defects appears to promote the reverse transformation of ferrite to austenite, since austenite has a higher solubility for carbon than ferrite [22, 33].

5.3.4 Role of the Cementite Morphology

Many investigations indicate that the cementite morphology (lamellar or spherical, which is often called spheroidite) and degree of its refinement influence strongly the process of strain-induced cementite dissolution. Shabashov et al. [16] investigated the microstructure and phase composition of pearlitic steel with different initial cementite morphology after severe plastic deformation by HPT using TEM and Mössbauer spectroscopy. They found that in case of spheroidite, more than 40% of the initial Fe_3C retained in the structure even after ten turns of HPT deformation, which corresponds to a shear strain, $\gamma = 600$. At the same time, for lamellar pearlite, the smaller the cementite lamellar thickness, the greater the degree of cementite dissolution. Similar results have been obtained for cold-rolled pearlitic steel samples [6]. It may also be noted that the size of pearlite colonies had no influence on the decomposition of cementite. Obviously, the larger inter-phase surface in fine pearlite compared with the one in coarse pearlite promotes more area for diffusion of C to ferrite thus accelerating the process of cementite dissolution. On the other hand, only fine lamellae with a thickness less than 30 nm can be deformed plastically [34], while thicker lamellae simply break into pieces as in Fig. 5.4a. The thinning and elongation of cementite platelets during straining lead to a further increase of the inter-phase boundary area, i.e., further facilitates the dissolution.

5.3.5 Driving Force and Mechanism of Strain Induced Decomposition of Cementite

There are two approaches to explain the dissolution of cementite. One approach is based on ballistic models where movement of matter is caused by external factors [1, 4, 35] and the other one is based on thermodynamic equilibrium considerations in metastable phases [7, 36–38]. Gridnev and Gavrilyuk [1] pointed out that the diffusion of carbon from cementite to the cores of the dislocations close to the interfaces is a possible mechanism for cementite dissolution. The bonding energy of carbon to the cores is 0.5 eV, which is approximately the same as the bonding energy to the cementite, so that carbon atoms may have low energy sites in the cores of dislocations in the ferrite matrix. Lojkowski et al. [3] and other authors [24, 34] indicated that with each cycle of plastic deformation of the material, the density of dislocations increases, in particular at the cementite–pearlite interface. Since diffusion along dislocation cores is orders of magnitude faster than in the bulk, this causes an increase of the effective diffusion coefficients and permits accelerated dissolution of cementite. At the same time, the degree of thermodynamic instability of the pearlite increases as well. This is caused by two reasons: their very small size so that capillarity effects become important [7, 39] and accumulation of strains in the precipitates [40].

The recent analysis of chemical composition of Fe_3C performed using APT revealed carbon loss in the carbide as a result of the severe plastic deformation. This offers a new insight into the reasons for the destabilization of cementite at plastic deformation. Off-stoichiometric cementite is not a stable phase [41–43]. The calculations of the densities of states of cementite having different carbon concentrations using the self-consistent unrestricted full-potential linear muffin-tin orbital method [43] have shown that the electron spectrum of cementite with a deficit of C atoms remains close to that of stoichiometric Fe_3C only until the C-vacancy concentration is about 25%, which corresponds to the absolute concentration of carbon in carbide of 19 at. %. Sharp changes in the electron spectrum were observed for the casses of larger concentrations of C vacancies, in these cases, the Fermi level was near the peak of the density of states, indicating the instability of such systems. The calculations of [43] demonstrate that the primary reason for the destabilization of cementite at plastic deformation is the deformation-induced outflow of carbon atoms from the cementite, and not capillary effects and relaxation of surface stresses as discussed by [7,27], because the partial decomposition of cementite observed after HPT deformation for $\gamma = 60$ had occurred when cementite lamellae still preserved something like their original shape (Fig. 5.2b) and their thickness was not small enough to consider the effect of capillarity.

On the other hand, further HPT deformation of pearlite leads to the thinning, elongation, and refinement of cementite platelets, and therefore, the contribution to the driving force from capillarity and Gibbs–Thompson effects as discussed by [27] starts to gain importance. As a result, the rate of cementite dissolution will increase with increasing shear strain leading to almost a complete dissolution. The APT data confirm the mechanism of outflow of carbon atoms from cementite lattice to dislocations in ferrite suggested by Gridnev and Gavrilyuk [1]. Actually, carbon segregations on dislocation cell walls close to the interphase boundary have been observed in APT experiments (Fig. 5.6). Frequent jumps of carbon atoms from cementite to dislocations would result in the formation of narrow areas in cementite along the interphase boundary depleted with carbon. However, the concentration of carbon gradually decreases by approaching this boundary as shown in concentration profile (Fig. 5.5b). This result indicates that there is downhill diffusion of carbon in cementite in direction to interphase boundary. The diffusion of carbon is facilitated by the formation of a nanocrystalline structure in the cementite (Fig. 5.4c) as grain boundary diffusion coefficients are always higher than for bulk diffusion. Therefore the outflow of carbon from cementite should be significantly accelerated subsequent to the formation of nanocrystalline structure.

5.4 Conclusions

1. Strain-induced decomposition of cementite occurs in pearlitic steels in conditions of severe plastic deformation. Decomposition of cementite begins with an outflow of carbon atoms from the carbides and induces the formation of substoichiometric cementite, which is thermodynamically unstable.
2. The morphology of carbides is very important to this process. Spheroidite and coarse lamellar pearlite dissolves much more slowly than fine pearlite deformed under the same conditions. Increased inter-phase area between ferrite and cementite in fine pearlite promotes better diffusion of C atoms; additionally, the thinning and elongation of cementite lamellae in fine pearlite due to plastic deformation further increases the inter-phase area.
3. Carbon atoms released after decomposition of cementite segregate to dislocations and grain and cell boundaries and do not seem to dissolve in the ferrite.

Acknowledgments The authors thank R.Z. Valiev and H. Rösner for their support for the experiments and fruitful discussions. The major part of this work was conducted during the fellowship of one of the authors (Yu. I.) supported by Alexander von Humboldt Foundation. Assistance from Messrs. W.A. Smith, A. Walker, and Miss N. Bielak with FIB specimen preparation is gratefully acknowledged.

References

1. V.N. Gridnev, V.G. Gavrilyuk, Phys. Metals **4**, 531 (1982)
2. Yu.V. Ivanisenko, G. Baumann, H.-J. Fecht, I.M. Safarov, A.V. Korznikov, R.Z. Valiev, Phys. Met. Metallogr. **3**, 303 (1997)
3. W. Lojkowski, M. Djahanbakhsh, G. Bürkle, S. Gierlotka, W. Zielinski, H.-J. Fecht, Mater. Sci. Eng. **A303**, 197 (2001)
4. Yu. Ivanisenko, W. Lojkowski, R.Z. Valiev, H.-J. Fecht, Acta Mater. **51**, 5555 (2003)
5. Yu. Ivanisenko, R. Wunderlich, R.Z. Valiev, H.-J. Fecht, Scripta Mater. **49**, 947 (2003)
6. W.J. Nam, Ch.M. Bae, S.J. Oh, S. Kwon, Scripta Mater. **42**, 457 (2000)
7. J. Languilaumme, G. Kapelski, B. Baudelet, Acta Mater. **45**, 1201 (1997)
8. K. Hono, M. Omuma, M. Murayama, S. Nishida, A. Yoshie, T. Takahashi, Scripta Mater. **44**, 977 (2001)
9. F. Danoix, D. Julien, X. Sauvage, J. Copreaux, Mater. Sci. Eng. **A250**, 8 (1998)
10. M. Umemoto, K. Todaka, K. Tsuchiya, Mat. Sci. Eng. **A375–377**, 899 (2004)
11. Z.G. Liu, X.J. Hao, K. Masuyama, K. Tsuchiya, M. Umemoto, S.M. Hao, Scripta Mater. **44**, 1775 (2001)
12. S. Ohsaki, K. Hono, H. Hidaka, S. Takaki, Scripta Mater. **52**, 271 (2005)
13. Y. Xu, M. Umemoto, K. Tsuchiya, Mater. Trans. **9**, 2205 (2002)
14. Z.G. Liu, H.J. Fecht, M. Umemoto, Mat. Sci. Eng. **A375–377**, 839 (2004)
15. A.V. Korznikov, Yu.V. Ivanisenko, D.V. Laptionok, I.M. Safarov, V.P. Pilyugin, R.Z. Valiev, Nanostructured Mater. **4**, 159 (1994)
16. V.A. Shabashov, L.G. Korshunov, A.G. Mukoseev, V.V. Sagaradze, A.V. Makarov, V.P. Pilyugin, S.I. Novikov, N.F. Vildanova, Mater. Sci. Eng. **A346**, 196 (2003)
17. D. Blavette, A. Bostel, J.M. Sarrau, B. Deconihout, A. Menand, Nature **363**, 432 (1993)
18. E. Bémont, A. Bostel, M. Bouet, G. Da Costa, S. Chambreland, B. Deconihout, K. Hono, Ultramicroscopy **95**, 231 (2003)

19. G. Da Costa, F. Vurpillot, A. Bostel, M. Bouet, B. Deconihout, Rev. Sci. Inst. **76**, 013304 (2005)
20. M.K. Miller, K.F. Russell, G.B. Thompson, Ultramicroscopy **102**, 287 (2005)
21. A. Menand, E. Cadel, C. Pareige, D. Blavette, Ultramicroscopy **78**, 63 (1999)
22. Yu. Ivanisenko, I. MacLaren, X. Sauvage, R.Z. Valiev, H.J. Fecht, Acta Mater. **54**, 1659 (2006)
23. R.Z. Valiev, Y.V. Ivanisenko, E.F. Rauch, B. Baudelet, Acta Mater. **44**, 4705 (1996)
24. I. MacLaren, Yu. Ivanisenko, H.-J. Fecht, X. Sauvage, R.Z. Valiev, In Y.T. Zhu, T.G. Langdon, Z. Horita, M.J. Zehetbauer, S.L. Semiatin, T.C. Lowe. [Eds.] Ultrafine Grained Materials IV. Proceedings. 2006 TMS Annual Meeting, San-Antonio, USA, March 18–21, 2006. TMS (The Minerals, Metals & Materials Society)
25. Yu. Ivanisenko, I. MacLaren, X. Sauvage, R.Z. Valiev, H.-J. Fecht, Sol. St. Phen. **114**, 133 (2006)
26. X. Sauvage, Yu. Ivanisenko, J. Mater. Sci. **42**, 1615 (2007)
27. X. Sauvage, J. Copreaux, F. Danoix, D. Blavette, Phil. Mag. A **80**, 781 (2000)
28. H.G. Read, W.T. Reynolds, K. Hono, T. Tarui, Scripta Mater **37**, 1221 (1997)
29. M.H. Hong, W.T. Reynolds, T. Tarui, K. Hono, Met. Mater. Trans. **30A**, 717 (1999)
30. X. Sauvage, N. Guelton, D. Blavette, Scripta Mater **46**, 459 (2002)
31. N. Maruyama, T. Tarui, H. Tashiro, Scripta Mater **46**, 599 (2002)
32. S. Ohsaki, K. Hono, H. Hidaka, S. Takaki, Scripta Mater. **52**, 271 (2005)
33. Yu. Ivanisenko, I. MacLaren, R.Z. Valiev, H.-J. Fecht, Adv. Eng. Mater. **7**, 1011 (2005)
34. G. Langford, Metal. Trans. **8A**, 861 (1977)
35. G. Martin, P. Bellon, Sol. State Phys. **50**, 189 (1996)
36. C. Suryanarayana, Progr. Mat. Sci. **40**, 1 (2001)
37. S.K. Pabi, J. Joardar, B.S. Murty, PINSA **1**, 1 (2001)
38. A.Y. Badmos, H.K.D.H. Bhadeshia, Metal. Trans. **28A**, 2189 (1997)
39. J. Weissmüller, J.W. Cahn, Acta Mater. **45**, 1899 (1997)
40. Yu.V. Ivanisenko, W. Lojkowski, R.Z. Valiev, H.-J. Fecht, Sol. State Phen. **94**, 45 (2003)
41. L. Battezzati, M. Baricco, S. Curiotto, Acta Mater. **53**, 1849 (2005)
42. F.X. Kayser, Y. Sumitomo, J. Phase Equilibria **18**, 458 (1997)
43. N.V. Medvedeva, L.E. Karkina, A.L. Ivanovskiy, Phys. Met. Metallogr. **96**, 16 (2003)

Chapter 6
Advanced Method for Gas-Cleaning from Submicron and Nanosize Aerosol

A. Bologa, H.-R. Paur, and H. Seifert

Abstract In a recent paper, the results of the development of an advanced method and an electrostatic precipitator of submicron and nanosize particles are presented. The method is based on particle charging in a DC corona discharge, formed in a small electrode gap of the electrode assembly and particle precipitation in the collector under the influence of aerosol space charge.

6.1 Introduction

The chemical process industry, electrical utility plants, waste incineration facilities, etc. require the efficient removal of fine and ultrafine particles from off-gases. Also the air pollution problem arise from the nanoparticles that are produced in industrial processes [1]. Dust is responsible for serious and disabling diseases like pneumoconiosis, interstitial lung disease and fibrosis, lung cancer, and asthma. The potential hazard of nanoparticles is not predictable by the bulk physicochemical properties of the different materials, but of the particle size, concentration, and toxicity. Thus, the removal of ultrafine particles from the exhaust gas is a new and important area of research [2].

The removal of submicron and nanosize particles from raw gases by cyclones, bag filters and scrubbers is a difficult task [3–5]. The collection of particles by cyclones causes considerable energy consumption. Fabric filters have problems in handling dust which may corrode or blind the cloth. Wet scrubbers could be used under favorable conditions However, their applications and performance are limited due to the highpower consumption and waste awter treatment

Electrostatic precipitators (ESP) are the most effective equipment for fine and ultrafine particle collection [6, 7]. The physical operation of a conventional ESP involves particle charging, collection, dislodging and disposal. Particles suspended in a gas enter the precipitator and pass through ionized zones. The high-voltage discharge electrodes through a corona effect emit negative ions into the gas with charged particles. The electric field around the highvoltage electrodes causes the charged particles to migrate and to precipitate on the collecting electrodes. Rappers dislodge the agglomerated particulate, which falls into the collection hoppers for removal [3, 5]. The greatest advantage provided by an ESP is that the electrostatic force of highly charged particles under the influence of an external electrostatic field is considerably large, compared with gravitational, thermal and inertial forces.

The efficiency of collection and the cost of an electrostatic precipitator depend on different factors: several of them are particle size and resistivity, operation voltage, gas flow distribution and proper operation of the collected aerosol handling system. The particle size of the incoming particulate has a dramatic impact on the sizing of an electrostatic precipitator and its cost. The size of a conventional precipitator must be increased in case of collection of very fine particles because they are easily reentrained into the gas stream. Particles in the lowrange resistivity (under $10^7 \, \Omega$ m) are easily charged. However upon contact with the collecting electrodes, they rapidly lose their charge and are repelled by the collecting electrodes back into the gas stream. Particles in the medium resistivity range $(10^7 - 10^{13} \, \Omega$ m) are the most acceptable for electrostatic precipitators. Particles in the high-resistivity category (above $10^{13} \, \Omega$ m, materials such as SiO_2, Al_2O_3, TiO_2, etc.) being collected may cause back corona on the surface of the grounded electrodes which decreases the collection efficiency of the electrostatic precipitator [7]. The proper operation of the highvoltage system is the key-factor of the electrostatic precipitator efficiency. Any problems or inefficiencies will be reflected in the operation of this system. The higher the operation voltage, the more complicated and costlier the highvoltage system. Good gas flow distribution is essential for optimum ESP operation. If the gas flow is not uniform through the precipitator, the ESP collection zone is not utilized effectively. The material collected by the electrostatic precipitator has to be removed at a greater rate than it is being collected. If not, the material will build up in the precipitator hoppers and eventually short out the highvoltage sections. If the material builds up into/between the collecting plates, the weight of the material can force the collecting electrodes apart and induce permanent damage in the electrodes.

Electrostatic precipitators suffer several phenomena when ESPs are used for cleaning gases with a high concentration of submicron and nanosize particles. This task is made especially difficult if particles are sticky (hydrocarbons, oil mists, tars) or when the gas stream is humid. Electrostatic precipitators suffer with the decrease of collection efficiency when dealing with corona discharge suppression. They also suffer with clogging problems due to buildup of fine particles on the high-voltage and collection electrodes leading to spark-over and a loss of efficiency.

The corona suppression phenomenon can be reduced and the collection efficiency of an electrostatic precipitator can be improved by a proper design of discharge electrode geometry, exhibiting a low corona onset voltage, by limiting the distance between the ionizing and collection stages and by a proper choice of conditions for particle collection.

6.2 Development of the Method and Electrostatic Precipitator

To ensure high collection efficiency for submicron and nanosize particles and to minimize the investment and operation costs, a new approach for electrostatic precipitation is realized in the electrostatic precipitator CAROLA® (Corona Aerosol

Abscheider), which is developed in the Forschungszentrum Karlsruhe [8–10]. The precipitator operates on the principle of unipolar particle charging in corona discharge and particle precipitation in the grounded collector under the external electric field and the field of aerosol own space charge.

In the CAROLA® precipitator, there is no need to impose the field by applying the high voltage in the collector stage, because the charge density associated with the aerosol is itself large enough for effective particle precipitation. The operation voltage of the CAROLA® precipitator ionizing stage is 10–20 kV. Thus, compared with conventional ESP, the investment and operation costs of the high-voltage part of the CAROLA® precipitator are reduced. The CAROLA® precipitator also differs from the conventional ESP by a high velocity of the gas flow in the ionizing stage.

The electrostatic precipitator consists of an ionizing and a collection stage (Fig. 6.1). Particles are charged in the DC corona discharge in the ionizing stage, which consists of high-voltage (HV) needle or star-shaped electrodes installed in the grounded nozzles, tube, or mesh electrodes (Fig. 6.2), depending on the CAROLA® application [6, 9, 10]. The high-voltage electrode is supported by a HV system connected to HV isolator. The electric gap is 10–15 mm.

In the electrostatic precipitator for oil mists (Fig. 6.1), the ionizing stage is installed inside of a collection stage, which together with grounded mesh electrode ensures the collection of charged particles.

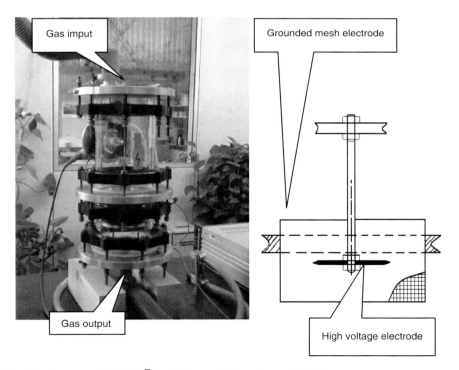

Fig. 6.1 One-stage CAROLA® precipitator and the schema of ionizing stage

a corona discharge at the top of the needle high voltage electrodes

b positive (left) and negative (right) DC corona discharge from the star-shaped high voltage electrode

Fig. 6.2 DC corona discharge in a small electrode gap. (**a**) Corona discharge at the top of the needle high voltage electrodes. (**b**) Positive (*left*) and negative (*right*) DC corona discharge from the star-shaped high voltage electrode

In the case of liquid aerosol, such water-based solutions of H_2SO_4 or HCl, the collector stage consists of a tube bundle or tower packing column. The collector stage is positioned downstream the ionizing stage without any plenum chamber [12].

In the case of solid aerosol, a bed-filter can be used for the precipitation of charged particles [13].

6.3 Influence of Gas Temperature on Current–Voltage Characteristics

The current–voltage characteristics are measured for air temperatures $T = 20 - 170°C$. The characteristics of a single-stage electrostatic precipitator with small electrode gap and star-shaped HV electrode are presented in Fig. 6.3.

With the increasing air temperature, the spark-over voltage linearly decreases; the corona current at operation voltage lower than the spark-over voltage linearly increases, but the spark over corona current remains practically constant.

6.4 Precipitation of Al_2O_3 Particles

A pilot one stage CAROLA® precipitator was used for the collection of Al_2O_3 particles from hot gases with a flow rate of 40–200 m^3/h. The temperature of the raw gas was 80–120°C with particle mass concentration up to 4 g/m^3. The ionizing

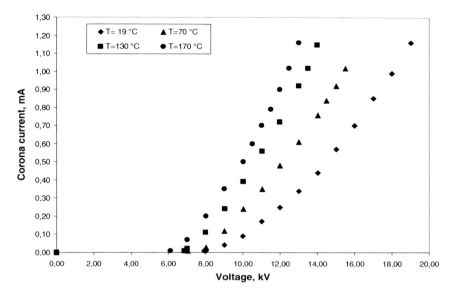

Fig. 6.3 Current–voltage characteristics of a single ionizing stage, negative DC corona

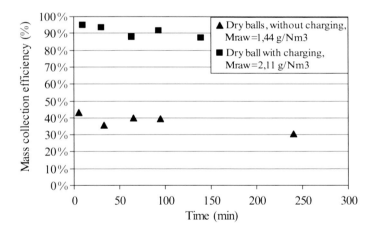

Fig. 6.4 Mass collection efficiency of the pilot precipitator, collection section with dry spheres, flow rate $100\,\text{m}^3/\text{h}$

stage consisted of 18 grounded nozzles (Fig. 6.2a) with needle high-voltage electrodes. Particles were collected in the collection stage with consisted of two beds of spheres with a diameter of 20–25 mm. The power consumption for particle charging was $\sim$200 W h/1, 000 m^3 of gas. The pressure drop in the collector was $P < 150$ Pa.

The data for mass collection $\eta_m = (1 - M_{\text{raw}}/M_{\text{clean}})\,100\%$ are presented in Fig 6.4. Here M_{raw} and M_{clean} are particle mass concentration in the raw gas and in the clean gas. Particle charging increases the mass collection efficiency of the

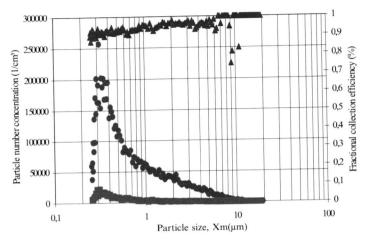

Fig. 6.5 Particle number concentration in the raw and clean gas and fractional collection efficiency of the pilot precipitator, two bed collector, $U = 12\,\text{kV}$, $I = 1.5\,\text{mA}$, flow rate $100\,\text{m}^3/\text{h}$, 18 nozzles, $t = 20\,\text{min}$, $M_{\text{raw}} = 2.7\,\text{g/Nm}^3$

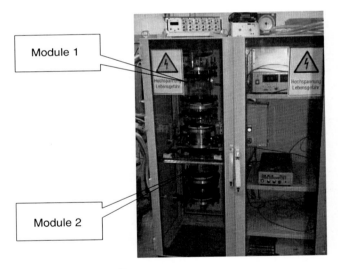

Fig. 6.6 Pilot two-module CAROLA® electrostatic precipitator

precipitator. The mass collection efficiency was $\eta_m > 90\%$ during the two hours of operation without cleaning the bed collector.

The fractional collection efficiency of the CAROLA® precipitator $\eta_f = (1 - C_{\text{raw}} - C_{\text{clean}})100\%$ was >99% for particles >10°μm; 93–98% for particles with sizes 1–6°μm; and 85–93% for submicron particles with size <1°μm (Fig. 6.5). Here C_{in} and C_{out} are particle fractional number concentrations upstream and downstream the precipitator.

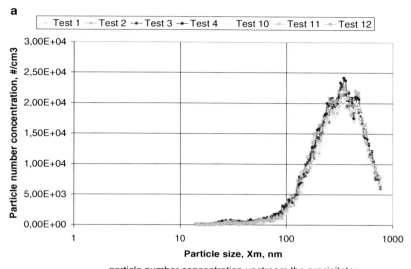

Fig. 6.7 Particle number concentration upstream (**a**) and downstream (**b**) the CAROLA® electrostatic precipitator, air flow rate 2 l/s. (a) Particle number concentration upstream the precipitator. (b) Particle number concentration downstream the precipitator

6.5 Precipitation of TiO$_2$ Particles

A two-modular CAROLA® electrostatic precipitator (Fig. 6.6) was used for the collection of submicron and nanosize TiO$_2$ particles.

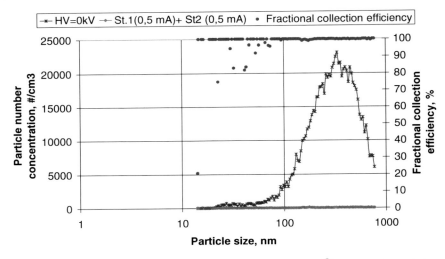

Fig. 6.8 Fractional collection efficiency of the two-module CAROLA® electrostatic precipitator, air flow rate 2 l/s

The air flow (flow rate 1.5–3.1 l/s, temperature 18–28°C) with TiO_2 particles was introduced into the precipitator. Particle number concentration varied from 0.2×10^5 to $7.0 \times 10^5 \#/cm^3$, depending on the setup operation parameters. The DC negative corona discharge was used for particle charging. Operation parameters of every module were: voltage $U = 15 \pm 0.5$ kV and corona current $I = 0.5 \pm 0.1$ mA.

Particle number concentrations upstream and downstream the electrostatic precipitator were measured by SMPS (Fa.TSI). The results of measurements are presented in Fig. 6.7. The use of CAROLA® precipitator ensures high fractional collection efficiency in the size range of 20–800 nm (Fig. 6.8).

6.6 Conclusion

A new method and an electrostatic precipitator CAROLA® are developed to control submicron and nanosized particles in the off gases. The precipitator operates on the principle of unipolar particle charging in corona discharge and particle precipitation in the grounded collector under the external electric field and the field of aerosol own space charge.

Single- and two-module pilot CAROLA® precipitators were tested for the collection of Al_2O_3 and TiO_2 particles at different gas temperatures, voltage up to $U \sim 15$ kV, and power consumption for particle charging up to 200 W h/$1{,}000$ m^3 of gas. The pilot unit ensures, for Al_2O_3 particle, a mass collection efficiency of up to 97% for raw gas with a particle mass concentration of up to 3 g/N m^3. The fractional particle number collection efficiency is in the range 85–93% for particles with sizes <1°μm and 93–100% for particles with sizes >1°μm. For TiO_2 aerosol, the use of CAROLA® precipitator allows to ensure a fractional collection efficiency of more than 99% for a particle size range of 20–800 nm.

Acknowledgments The authors thank Dipl.-Ing. Th. Wäscher and Mr. K.Woletz for their skilled technical works. The authors gratefully acknowledge the support from the Institute für Reaktorsicherheit, FZK and Fa. Almatis GmbH in the framework of cooperation projects.

References

1. M. Li, P.D. Christofides, Ind. Eng. Chem. Res. **45**, 8484 (2006)
2. Y. Zhuang, Y.J. Kim, T.G. Lee, P. Biswas, J. Electrostat. **48**, 245 (2000)
3. E. Weber, W. Brocke, *Apparate und Verfahren der industriellen Gasreinigung, Band1* (Feststoffabscheidung, Oldenburg Verlag, 1973)
4. W. Light, *Air Pollution Control Engineering: Basic Calculations for Particulate Collection* (Marcel Dekker, Inc., New York and Basel, 1988)
5. F. Löffler, *Staubabscheidung* (Georg Thieme Verlag, Stuttgart, 1988)
6. J.R. Melcher, K.S. Sachar, E.P. Warren, Proc. IEEE **65**(12), 1659 (1977)
7. K.R. Parker ed. *Applied Electrostatic Precipitation* (Blackie Academic & Professional, 1997)
8. A.M. Bologa, H.-R. Paur, Th. Wäscher, W. Baumann, DE Patent N 10132582
9. A.M. Bologa, H.-R. Paur, H. Seifert. Th. Wäscher, IEEE Trans. Industry Applic. **41**(4), 882 (2005)
10. A. Bologa, H.-R. Paur, H. Seifert, K. Woletz, in *Proceedings of the International Conference & Exhibition for Filtration and Separation technology, FILTECH 2007*, Wiesbaden, Germany, 2007
11. A. Bologa, Th. Wäscher, H.-R. Paur, R. Arheidt, DE Patent N 10 2005 023521
12. A. Bologa, H.-R. Paur, K. Woletz, DE Patent N 10 2006 055543
13. A. Bologa, R. Arheidt, H.-R. Paur, H. Seifert, W. Lingenberg, G. Weber, Chemie Ingenieur Technik **75**(8), 1058 (2003)

Chapter 7
Deformation Microstructures Near Vickers Indentations in SNO$_2$/SI Coated Systems

G. Daria, H. Evghenii, S. Olga, D. Zinaida, M. Iana, and Z. Victor

Abstract The micromechanical properties (hardness and brittleness) of the hard-on-hard SnO$_2$/Si-coated system (CS) and their modification depending the on load value has been studied. A nonmonotonic changing of microhardness with load growth was detected. The brittle/plastic behavior of the rigid/hard-on-hard SnO$_2$/Si CS and its response to concentrated load action explains it.

A specific evolution of the indentation-deformed zone vs. load value attributed to the change in the internal stress redistribution between film and substrate was detected. It results in a brittleness indentation size effect (BISE) of the SnO$_2$/Si CS revealed in this experiment.

It was shown that the greater portion of internal stresses under indentation is concentrated in the coating layer at small loads. This fact causes a strong elastic–plastic relaxation in the film and its delamination from substrate. The increase of brittle failure in the indentation-deformed zone with a decrease of indentation load was revealed.

7.1 Introduction

The photovoltaic method of the direct conversion from solar energy into electrical one has found an increasing application in the last years as environmentally safe method for renewable resources of *electricity*. There are currently a number of technologies and semiconductor materials under investigation under conditions of continuous search for the ways to increase the energy conversion efficiency and to reduce the price of the solar cells. Examples include the first installations of solar batteries on the basis of ITO/Si and SnO$_2$/Si that have been already obtained and applied in practice. A new low-cost promising technology for the fabrication of SnO$_2$/Si plates for solar cells was elaborated and described in work [1]. Due to pronounced metallic conductance and highest transparency for solar beams, the SnO$_2$ film is an ideal material for the conversion of solar radiation into electrical energy. At the same time, an obvious demand arises to develop the fabrication technology of these planar structures with optimal photovoltaic and mechanical properties. It is known that the durability and long life of new structural materials and devices on their base depend on the microstructural state that influences in turn on such mechanical parameters as microhardness (H), fragility (γ), adhesion degree, etc. From this point of view, the determination of the micromechanical properties and the identification of factors that influence them are an important part of investigations for obtaining high-quality solar cells.

Note that the aforementioned materials represent the so-called coated system (CS), namely, thin films coating on the tridimensional substrates. A minimization of film thickness in CS to submicronic or even nanometric dimensions leads to a significant, sometimes unpredictable, modification of the mechanical properties of materials [2–6].

Therefore, the work program assumes the study of microstructure and the examination of the strength and brittleness parameters of a composite structure of the "hard-on-hard" type, namely SnO_2/Si. The obtained results will expand the knowledge of deformation peculiarity and will contribute to the optimization of fabrication of these structures.

7.2 Experimental

To consider the specificity of deformation of the SnO_2/Si composite materials under concentration load action, the Vickers quasi-static microindentation method was used in the work. The loading value (P) varied in the range of 0.3–1.0 N for different specimens. In all experiments, the indentation depth exceeded the film thickness ($h_{ind} > t$). The microhardness was estimated by the standard formula: $H = 1,854 P/d^2$, where P is the load and d is the indentation diagonal [7]. The d size was appreciated as an average of both the diagonals of ten imprints. The microstructural examinations have been performed by means of Vega Tescan scanning electron microscopy (SEM), Amplival, Neophot, and XJL-101 light microscopy (LM). All the researches were performed at room temperature (RT). The thickness of the SnO_2 coatings for experiments varied within $t = 350$–400 nm.

7.3 Results and Discussion

The microhardness changes depending on the applied load are shown in Fig. 7.1 for the SnO_2/Si-coated system and for the "pure" Si single-crystal used as a substrate for the investigated CS. One can note a difference in the shape of the curves. For the SnO_2/Si CS, in general, an increase of the microhardness with loading growth is observed (Fig. 7.1a), whereas the curve for pure Si demonstrates a decrease with growth of loading in the low load interval, which is stabilized in the region of heavier loads (Fig. 7.1b).

The behavior of $H(P)$ curve for the pure Si is characteristic of the hard-rigid monocrystalline solids. As it follows from literature data [7, 8], there are several

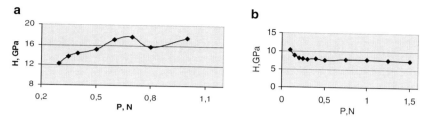

Fig. 7.1 The microhardness-vs.-load dependence for the SnO_2/Si CS (**a**) and Si substrate (**b**); Si plane (001); thickness of SnO_2 film is ≈ 350 nm, indentation is made at $T = 300$ K

causes for such a shape of Si curve. First, it is connected with the elastic recovery of indentations, which is more pronounced at small load. The indent elastic recovery leads to H increase. Second, a brittle destruction appears near indentations beginning from certain load and it enlarges further with load increase. The brittle destruction of indentation leads to the hardness diminution [7–9]. As for the SnO_2/Si curve, its course is rather unusual: a nonmonotonic modification of microhardness values with load growth can be observed in Fig. 7.1a. To understand the specific character of these changes and the curve course as a whole, we have carried out an examination of the indentation shape and indentation surroundings (Fig. 7.2).

Figure 7.2 shows that the film surface microstructure has a globular character. Comparing the background relief of film surface (Fig. 7.2a, c, e) with the image of the indenter-sample contact surface (Fig. 7.2b, d, e), we observe that the globule dimensions are basically similar for both the background and the indentations. The difference consists in their diversiform. The surface inside the indents is sufficiently plastic and the globules have assumed a pressed shape, whereas those of the background surface look as though having a spherical one (to compare Fig. 7.2a, c, e with Fig. 7.2b, d, f).

In addition, a distance between some of globules inside the indents became greater compared with the distance between globules on the film surface outside the indentations. This effect is a consequence of the indenter action on the studied CS. We suppose the globules flattening in the contact zone with partial plastic flow and material outlet from under the indenter during penetration. As a result, the film thickness becomes less. This supposition is in accordance with Fig. 7.2 of the work [10] where the diminution of layer thickness in the indentation contact zone of GaAs–AlAs heterostructures can also be observed. The spectral analysis made for both the background and contact zones demonstrates the thickness diminution of the SnO_2 layer in the SnO_2/Si CS as a response to the indenter action (pressing). For example, the spectrograms illustrating the content of elements in the sample inside and outside of the indentation made at $P = 0.3$ N are presented in Fig. 7.3.

One can see the presence of five elements: Si, being a substrate material, Sn and O being the basic ingredients creating the SnO_2 film, and C and N being the insignificant attendant impurities. A diminution of the percentage of coating elements inside the indentation in comparison with those of the outside region takes place

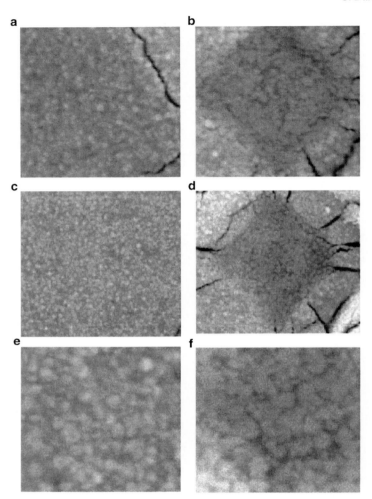

Fig. 7.2 SEM, SnO_2/Si CS. The microstructure of film surfaces outside the indentations (**a, c, e**) and microstructure inside ones (**b, d, f**). P (N): 0.3 (**a, b**), 0.5 (**c, d**), 1.0 (**e, f**)

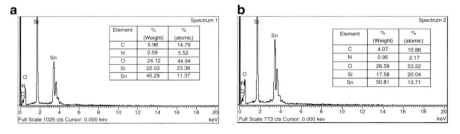

Fig. 7.3 SEM, EDX spectroscopy. The chemical composition of the SnO_2/Si CS inside a Vickers indentation (**a**) and of the film in the indentation surrounding (**b**) is shown

7 Deformation Microstructures Near Vickers Indentations

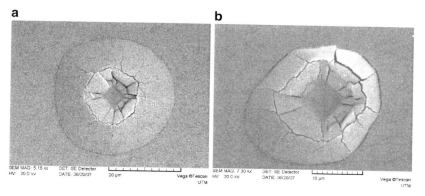

Fig. 7.4 SEM, the Vickers method. The shape of brittle/plastic deformed zones in the vicinity of indentations made on the SnO_2/Si composite structure at loading of 0.3 N (**a**) and 0.5 N (**b**)

Table 7.1 Relation between the dimension of indentation diagonals and brittle/plastic deformed zones revealed on the SnO_2/Si composite structure as a response to the Vickers indenter penetration

	SnO_2/Si CS		
$P(N)$	$d(\mu m)$	$D_f(\mu m)$	$\gamma_{cs} = D_f/d$
0.3	6.2	34	5.48
0.4	7.5	26	3.47
0.5	8.3	26	3.13
1.0	12.0	26	2.17

for the studied CS (45.29% and 21.12% against 50.81% and 26.59% for Sn and O respectively). This result is an evidence of plastic flow and material movement during indenter penetration.

However, although the contact zone between CS and indenter was shown to be plastic, the indentation neighborhoods demonstrated a significant brittle destruction of the SnO_2 coating. For instance, the SEM images of two indentations, plotted at $P = 0.3$ and 0.5 N, and the destructed zone around them are presented in Fig. 7.4. It is clearly seen from these pictures that the diameter of the destruction zone (D) is much larger than the indentation diagonal (d).

An indentation size effect (ISE) of brittleness was revealed on the SnO_2/Si CS for indentations performed at various loads: (P, N): 0.3; 0.4; 0.5; and 1.0. It was established that the relation between the size of destroyed film surface (D_f) and the indentation diagonal (d) is a function of the loading value, or otherwise, they are dependent on the indentation depth. Thus, an unusual behavior of the fracture character in the indentation neighborhood has been detected with the increase of the applied load (Table 7.1).

As it follows from Fig. 7.4 and Table 7.1, the diameter of the destroyed zone of film near indentations (D_f) made on the SnO_2/Si CS at different loads decreases with the applied load increase, contrary to expectations, from 34 to 26 μm for 0.3 and 1.0 N respectively. The effect became more appreciable when considering the dimensionless parameter $\gamma_{CS} = D_f/d$. In that case, γ_{CS} parameter decreased from

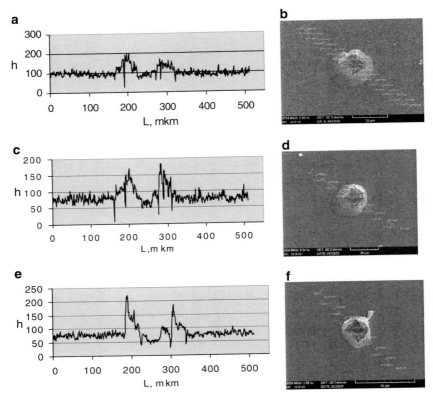

Fig. 7.5 SEM, SnO_2/Si. The cross-sectional view of surface of indents with surroundings (**a, c, e**) made at three loads (P, N): 0.4 (**a, b**), 0.5 (**c, d**), 1.0 (**e, f**) and respective images full face (**b, d, f**). Axes of coordinates: X (μm); Y, nm

5.48 to 2.17 for the 0.3 and 1.0 N accordingly. To explain this ISE of brittleness, we will refer to surface relief in the indentation-deformed zones.

The comparison of the surface profile around indentations made on the SnO_2/Si CS at different loads (Fig. 7.5) demonstrated that the intensive pileups are created around indentations. Most probably, two processes cause them: (1) plastic–elastic recovery of indentation and the appearing of pileups in Si substrate [7, 11]; (2) the SnO_2 film delamination from the substrate.

As it follows from Fig. 7.5, the area extent of the pile-up zone is approximately equal for all loads, whereas the indentation dimension visibly grows both in diameter and in depth with the increase of load.

This evolution of the indentation-deformed zone vs. load can be attributed to the change in the internal stress redistribution between film and substrate. When load value is small ($P = 0.3$ N), the indentation depth (h_{ind}) is about 0.9 μm and proportion t/h_{ind} is $\sim (0.4{:}0.6)$. For larger loads, namely $P = 1.0$ N, h_{ind} became ~ 1.7 μm and $t/h_{ind} \approx 0.2 : 0.8$. Hence, in the first case, the internal stresses will be distributed approximately equally between the film and substrate volume, whereas in the second one, the main part of stresses will affect the substrate volume, 80% and 20% of stresses being distributed in substrate and film accordingly. As a result, at low loads, the strain is preferably distributed in film volume (in width); at heavy loads, in substrate volume (in depth). It means that the heavier the loading, the greater the deformation part in the substrate volume compared with the film one; i.e., the destruction region on the CS surface layer is proportionally diminished. In turn, the diminution of the brittle fracture around indentation leads to hardness increase. Note that the redistribution of stresses between film and substrate volumes not always occurs monotonically at load growth. As a result, the dependence hardness vs load has a nonmonotonic character. Thus, the revealed brittle/plastic behavior of the rigid/hard-on-hard SnO_2/Si CS studied in the work and its mechanical response to concentrated load action explain the nonmonotonic character of the $H(P)$ curve and the increase of microhardness value vs. load one (see Fig. 7.1a).

One more aspect should be taken into consideration (see Fig. 7.5a, c, e). The surface inside the indentation is smoother for a load of 1.0 N compared with 0.4 and 0.5 N. This fact confirms the aforementioned suggestion that the globule flattening occurs in the contact zone and the effect becomes more visible at heavier load (see Fig. 7.2).

In addition, for a rigid/hard CS such as SnO_2/Si, it was demonstrated that a great portion of brittle failure is concentrated in the film volume and a smaller part is distributed in the substrate volume.

7.4 Conclusions

The study of micromechanical properties, namely, microhardness (H), brittleness (γ), and adhesion degree of the hard-on-hard SnO_2/Si CS and modification of these parameters depending on the load value has been carried out. A nonmonotonic insrease of microhardness with load growth has been observed. The brittle/plastic behavior of the rigid/hard-on-hard SnO_2/Si CS and its mechanical response to concentrated load action explained this result.

A specific evolution of the indentation-deformed zone vs. load value attributed to the change in the internal stress redistribution between film and substrate was detected. It results in the brittleness indentation size effect (BISE) of the SnO_2/Si CS revealed in this experiment.

It was shown that the greater portion of internal stresses under indentation is concentrated in the coating layer at small loads. This causes strong elastic–plastic relaxation in the film and its delamination from substrate. Larger brittle destruction in the indentation-deformed zone for lower loads compared with heavier ones was shown.

Acknowledgments The reported work was funded by the Supreme Council for Science and Technological Development of the Academy of Sciences, Republic of Moldova. The research was partially performed with the support of the National Center of Material Science of the Technical University of Moldova (grant RESC-MR-995).

References

1. A. Simashkevich, D. Sherban, L. Bruc, A. Coval, V. Fedorov, E. Bobeico, Iu. Usatyi, in *Proceedings of the 20th European PV Solar Energy Conference*, Barcelona, 2005, p. 980
2. W.A. Soer, J.Th.M. De Hosson, A.M. Minor Jr., W.J.W. Morris, E.A. Stach, Acta Mater. **52**, 5783 (2004)
3. G. Golan, E. Rabinovich, A. Axelevitch, A. Seidman, N. Croitoru, J. Optoelectron. Adv. Mater. **2**, 317 (2000)
4. S.P. Moylan, S. Kompella, S. Chandrasekar, T.N. Farris, J. Manufact. Sci. Eng. **125**, 310 (2003)
5. V.Ya. Malakhov, Math. Phys. Chem. **102**, 291 (2002)
6. M.L. Trunov, J. Phys. D Appl. Phys. **41**, 074011 (2008)
7. S. Yu, D.Z. Boyarskaya, X. Grabco, M.S. Kats, Kishinev **294**, (1986) (in Rus.)
8. M.I. Val'kovskaya, B.M. Pushkash, E.E. Maronchiuk, Kishinev **107**, (1984) (in Rus.)
9. Yu.S. Boyarskaya, D.Z. Grabko, M.S. Kats, J. Mater. Sci. **25**, 3611 (1990)
10. M.R. Castell, G. Shafirstein, D.A. Ritchie, Philos. Mag. A **74**, 1185 (1996)
11. O. Shikimaka, PhD Thesis in physics and mathematics, www. cnaa. acad. md, 2005 (in Romanian)

Chapter 8
Grain Boundary Phase Transformations in Nanostructured Conducting Oxides

B.B. Straumal, A.A. Myatiev, P.B. Straumal, and A.A. Mazilkin

Abstract Nanostructured conducting oxides are very promising for various applications like varistors (doped zinc oxide), electrolytes for the solid oxide fuel cells (SOFC) (ceria, zirconia, yttria), semipermeable membranes, and sensors (perovskite-type oxides). Grain boundary (GB) phases crucially determine the properties of nanograined oxides. GB phase transformations (wetting, prewetting, pseudopartial wetting) proceed in the conducting oxides. Novel GB lines appear in the conventional bulk phase diagrams. They can be used for the tailoring of properties of nanograined conducting oxides, particularly by using the novel synthesis method of liquid ceramics.

8.1 Introduction

Conducting oxides are currently broadly used for various applications, e.g., zinc oxide for manufacturing of varistors [1, 2], ruthenates as thick-film resistors [3], oxides of fluorite structure (ceria, zirconia, yttria) as electrolytes for the solid oxide fuel cells (SOFC) and oxygen sensors [4], perovskite-type oxides ($BaTiO_3$, $SrTiO_3$, $LaAlO_3$, $LaCrO_3$, etc.) as electrolytes and electrodes for SOFC, semipermeable membranes, and sensors [5]. Other applications of semiconducting oxides are various electronic devices such as self-controlled heaters, color TV degaussers, fuel evaporators, and air-conditioning equipment [6].

The electrical properties of these oxides, especially of nanostructured ones, are crucially determined by the structural and chemical characteristics of the grain boundaries (GBs). It can be due to the formation of: (a) conventional GB segregation layer with a content of a second (third, fourth etc.) component less than one monolayer (ML); (b) thin (few nm) continuos layer of a GB phase, which can be described also as multilayer segregation, and (c) thick (several micrometers and more) layer of a solid, liquid, or amorphous wetting phase. Such GB layers may be thermodynamically stable, metastable, or unstable. Therefore, it is of crucial importance, to have at disposal the phase diagrams including the lines of bulk and GB phase transformations. Such diagrams allow tailoring the synthesis of nanostructured oxides, controlling their microstructure and producing the devices with stable properties and long life-time.

8.2 Grain Boundary Phase Transformations and Phase Diagrams

Let us consider the schematic two-component eutectic phase diagram describing the conditions for the thermodynamic equilibrium for all three cases listed earlier. Thermodynamically stable GB layers form as a result of the so-called GB phase transitions, GB wetting being an important example of such processes [7, 8]. GB wetting phase transitions have recently been included in the traditional phase diagrams of several systems [9, 10]. The occurrence of wetting depends on the GB energy, σ_{GB}. Consider the contact angle Θ between a bicrystal and a liquid phase. When σ_{GB} is lower than $2\sigma_{SL}$, where σ_{SL} is the energy of the solid–liquid interphase boundary, the GB is nonwetted and $\Theta > 0°$ (point 1 in Fig. 8.1). However, if $\sigma_{GB} \geq 2\sigma_{SL}$, the GB is wetted and the contact angle $\Theta = 0°$ (point 2). The temperature dependency of $2\sigma_{SL}$ is stronger than that of σ_{GB}. If the curves describing the temperature dependencies of σ_{GB} and $2\sigma_{SL}$ intersect, the GB wetting phase transition will occur upon heating at the temperature, T_w, of their intersection. At $T \geq T_w$, the contact angle is $\Theta = 0°$. By crossing the bulk solidus between points 2 and 3, the liquid phase becomes metastable. Its appearance in the system costs the energy loss Δg. The energy gain $(\sigma_{GB} - 2\sigma_{SL})$ above T_w can stabilize the GB liquid-like layer of a thickness l. By moving from point 3 to point 4, the energy loss Δg increases and the GB liquid-like layer disappears at GB solidus line. Therefore, the stable layer of liquid-like phase (which is unstable in the bulk) can exist in the GB between bulk and GB solidus lines (point 3). The same is true also if the second phase is solid. In point 2′ GB in the α-phase has to be substituted by the layer of

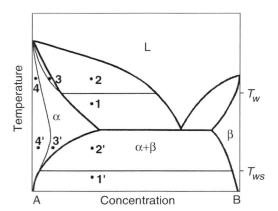

Fig. 8.1 Schematic phase diagram with lines of GB phase transitions. T_w – temperature of the GB wetting phase transition (proceeds between points 1 and 2). T_{ws} – temperature of the GB solid phase wetting transition (proceeds between points 1′ and 2′). Between points 3 and 4 the GB premelting phase transition occurs. Between points 3′ and 4′ the GB premelting phase transition occurs. In points 3 and 3′ GB is covered by the equilibrium layer of a liquid-like or β-like phase which is unstable in the bulk

β-phase and two α/β interphase boundaries (IBs). In point 3′ GB is covered by the equilibrium layer of a β-like phase that is unstable in the bulk. In points 4 and 4′ GB is "pure" and contains only the usual segregation layer of component B. Therefore:

(a) Conventional GB segregation layer with a content of a second (third, fourth etc.) component less than one ML exists in areas marked by points 4 and 4′;
(b) Thin (few nanometres) continuous layer of a GB phase exists in areas marked by points 3 and 3′ and
(c) Thick (several micrometres and more) continuous GB wetting layer of a liquid or solid phase exists in areas marked by points 2 and 2′.

This simple scheme permits to understand the phenomena in numerous conducting oxides. Very often, they are produced with the aid of the liquid phase sintering, where all GBs are wetted by liquid phase (i.e., in the area 2 of the scheme in Fig. 8.1). By the following cooling, the GB melt layer solidifies and can transform either into an array of droplets, or into amorphous GB layer, or into crystal wetting phase, or into conventional GB segregation layer of less than one ML. In detail, the GB phases and GB structure in conducting oxides determining their life-time and properties strongly depend on the composition and the processing route.

8.3 Grain Boundary Phases in Zinc Oxide

Zn oxide is mainly used for manufacturing varistors. Varistors exhibit highly nonlinear current–voltage characteristics with a high resistivity below a threshold electric field, becoming conductive when this field is exceeded, enabling them to be used in current over-surge protection circuits [11]. The model usually proposed to account for the electrical properties of ZnO-based varistors is constituted on the basis of a bricklayer. ZnO-based varistors are approximated as a stacking of good conducting grains separated by GBs, which support back-to-back double Schottky barriers [12–14]. Polycrystalline zinc oxide contains small amounts of dopants, mainly bismuth oxide. After liquid-phase sintering, such material consists of ZnO grains separated by the Bi_2O_3-rich GB layers. Interfaces between the ZnO grains control the nonlinear current–voltage characteristics. Although the Schottky barriers at ZnO/ZnO boundaries mainly control the voltage-dependent resistivity of a varistor, the Bi-rich GB phase also inputs into the overall resistivity.

The intergranular phase originates from the liquid-phase sintering. The sintering conditions alter the performances of ZnO varistors [13]. An increase in the sintering temperature results usually in a lowering in the nonlinearity of the current–voltage curve. Bhushan et al. pointed out that an increase in the sintering temperature would lower the Schottky barrier height [15] and Wong mentioned that the volatilization of Bi_2O_3 during the sintering would bring a loss in the nonohmic property of the varistors [16]. The big amount of structural investigations permitted us to construct the GB lines in the ZnO–Bi_2O_3 bulk phase diagram (Fig. 8.2) [1, 17–25]. The first variant of the ZnO–Bi_2O_3 phase diagram has been experimentally constructed by

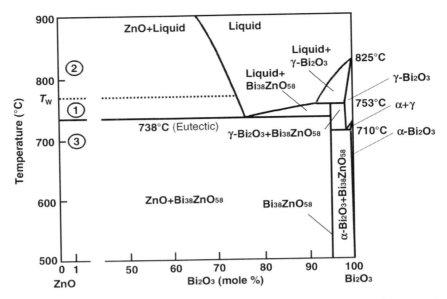

Fig. 8.2 ZnO–Bi$_2$O$_3$ phase diagram (*solid lines*, [27]). Tie-line of GB wetting phase transition slightly at T_w above eutectic temperature T_e is added to the bulk diagram (*dotted line*). In the area (1) between T_e and T_w melt partially wets the ZnO GBs [1]. In the area (2) above T_w melt fully wets the ZnO GBs [5]. In the area (3) below T_e ZnO GB triple junctions contain crystalline Bi$_2$O$_3$ and ZnO GBs contain amorphous Bi-rich phase with about 25–30 mol.% Bi (see scheme in Fig. 8.3b) [1, 17]

Safronov et al. [26]. However, recently Guha et al. [27] found new γ-Bi$_2$O$_3$-phase and refined the ZnO–Bi$_2$O$_3$ phase diagram (Fig. 8.2).

The liquid phase sintering of the ZnO + Bi$_2$O$_3$ mixture proceeds in the ZnO + liquid region of the ZnO–Bi$_2$O$_3$ phase diagram, i.e., above eutectic temperature of $T_e = 738°C$ (usually at 850°C) [1]. During the liquid phase sintering, all ZnO/ZnO GBs are completely wetted by the thick layer of the melt. The thickness of the melt layer is governed only by the grain size and amount of the liquid phase (i.e., on the Bi$_2$O$_3$ content). At 850°C, the liquid phase completely wets not only all ZnO/ZnO GBs, but also the free surface of the ZnO particles [17]. There is some indications that in the ZnO + liquid region close to T_e, the complete GB wetting transforms into partial GB wetting (with contact angles above zero) [1]. In other words, in the ZnO–Bi$_2$O$_3$ phase diagram, the GB wetting tie-line exists slightly above T_e (Fig. 8.2).

The quenching from 850°C leaves a thick intergranular phase at the ZnO/ZnO GBs. However, the slow cooling below T_e leads to the dewetting of ZnO/ZnO GBs by crystallization of Bi$_2$O$_3$ [18–21]. Since the optimization of the varistor properties needs the slow cooling or a low-temperature post annealing, much work was devoted to the structure of GBs in varistors [20–22]. At the beginning of these investigations, it was believed that all GBs contain thin Bi-rich intergranular phase.

8 Grain Boundary Phase Transformations in Nanostructured Conducting Oxides

Then Clarke reported that most ZnO GBs in a commercial varistor were free from the second-phase films, and the atomically abrupt GBs were observed using the lattice fringe imaging [28]. However, later Olsson et al. found the continuous Bi-rich films in the majority of ZnO/ZnO GBs, and only a few GBs were atomically ordered up to the GB plane [23, 24]. It was also found that the treatment at high hydrostatic pressure of 1 GPa leads to the desegregation of ZnO/ZnO GBs [25]. During desegregation, the Bi-rich GB phase disappears due to the Bi GB diffusion toward the secondary phase in the GB triple junctions.

Wang and Chiang studied the ZnO with 0.23 mol% at Bi_2O_3 700°C [1]. The samples were brought into equilibrium at this temperature from three different starting points: (a) after liquid phase sintering at 850°C followed by 24 h annealing at 700°C and slow cooling down to the room temperature; (b) by sintering directly at 700°C (i.e., below T_e, without presence of any liquid phase) for 2 h by 1 GPa followed by the annealing at 700°C at the room pressure; and (c) equilibrium segregation at 700°C was reached from the high-pressure desegregated state. Wang and Chiang discovered that in all the three cases the equilibrium GB state at 700°C is the amorphous intergranular film of 1.0–1.5 nm in thickness. In other words, a thin intergranular film has a lower free energy in comparison with pure crystal–crystal GB. The thermodynamic conditions for the existence of such films were studied by Clarke [29]. After desegregation at high temperature (Fig. 8.3a), GBs are free from any Bi-rich layers (thin or thick). Crystalline Bi_2O_3 particles are present in the GB triple junctions. However, after additional annealing at the same temperature of 700°C but at atmospheric pressure, Bi diffuses back from the triple junctions into the GBs forming the amorphous GBs films of 1.0–1.5 nm in thickness (Fig. 8.3b). In other words, the amorphous films build not from the undercooled liquid, but in the solid phase, as a result of Bi GB diffusion. Moreover, the thin amorphous film covers not only the ZnO/ZnO GBs, but also the intrephase boundary between ZnO grains and Bi_2O_3 particle in the ZnO GB triple junction (Fig. 8.3b).

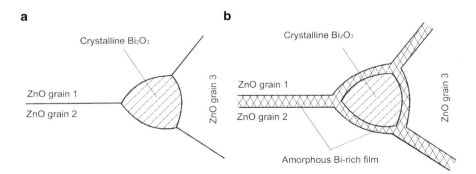

Fig. 8.3 Scheme of GBs and GB triple junction in the ZnO–Bi_2O_3 at 700°C [1]. (**a**) Structure after pressure desegregation at 1 GPa. GB triple junction contains lenticular crystalline Bi_2O_3 phase. GBs contain no films. (**b**) Structure after additional anneal at atmospheric pressure. GBs contain amorphous Bi_2O_3-rich film of 1–2 nm thickness with about 25–30 mol.% Bi. Similar film separates ZnO grains and the lenticular crystalline Bi_2O_3 phase in the GB ZnO triple junction

This behavior can be explained by the so-called pseudopartial wetting [17, 30]. At certain thermodynamic conditions, liquid droplets have a nonzero contact angle with a solid substrate (or a GB), but the rest of a substrate surface (or a GB) is not dry, but covered by a thin film of few nm thickness. For example, the liquid Bi-rich nanodroplets (5–15 nm) with contact angle of about 40° were observed on the top of the amorphous film of 1.95 nm thickness on the ZnO surface facets [17].

8.4 Conducting Oxides of Fluorite Structure

Conducting oxides of fluorite structure have received much attention in recent years due to their ionic conductivity with the applications as electrolytes for the SOFC and oxygen sensors. Yttria-stabilized zirconia (YSZ) is by far the most widely used solid electrolyte for technological applications. The main factors driving the interest for this solid electrolyte are its high chemical stability in oxidizing or reducing environments and its compatibility with a variety of adjoining electrode materials. It is presently employed at temperatures above 600°C. Other oxides like calcia or scandia can also be used for the stabilization of zirconia. Although stabilized zirconia exhibits good conductivity at high temperatures, the need for a better oxygen-conducting material in SOFCs has shifted interest to doped ceria [31], which exhibits good conductivity at lower temperatures. Usual doping ions for CeO_2 are Gd^{3+}, Sm^{3+}, and Y^{3+}. Substitution of the Ce^{4+} cations in the lattice results in the formation of vacancies and enhances the ionic conductivity.

8.4.1 GB Wetting Phases

It has been shown that the maximum of the ionic conductivity of YSZ occurs around 9.5 mol% Y_2O_3 [32,33]. Measurements of conductivity and oxygen diffusivity confirmed that YSZ are the ionic conductors at the temperatures as low as 200°C [34]. Critical to the low-temperature applications are the internal interface properties of YSZ. In YSZ, a glassy phase was commonly observed in GBs and GB triple junctions. In [38] two YSZs (called Z_C and Z_F) were sintered from powders prepared through two different processing routes. In samples Z_C the glassy phase wetted GBs and GB triple junctions. Glassy phase in triple junctions has a shape of stars with zero contact angles at GBs. These "stars" continue toward GBs as GB wetting layers. In samples Z_F, the amorphous precipitates of glassy phase in triple junctions are lenticular, and spherical glass pockets are widely dispersed in the bulk of grains, but there is no evidence of glassy films at GBs. As a result, the GB conductivity of the Z_F polycrystal, which shows glass-free GBs, is about three orders of magnitude higher than that of the Z_C material (Fig. 8.4). These results are consistent with the mechanism of oxygen-ion transport across GBs suggested by Badwal [35]. Conductivity occurs without any constriction of current pathways in the Z_F

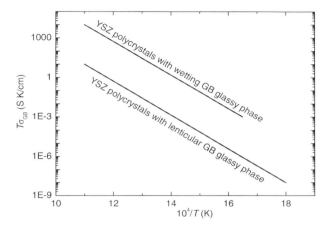

Fig. 8.4 Temperature dependence of specific GB conductivity of YSZ polycrystals with and without GB glassy phase according to the data of [34]

ceramics, while it is restricted to the unwetted GBs in the Z_C ceramics. Therefore, if a GB wetting phase is detrimental, one can change a composition in such a way that the GB wetting conditions are not fulfilled any more. In this case the GB network of detrimental phase is broken and the properties of a material improve. Thus, by changing GB wetting conditions by microalloying, one can improve the properties of a conducting oxide.

8.4.2 Monolayer GB Segregation

Even in the absence of the GB layers of wetting phases, the properties of conducting oxides can be controlled by the conventional (less that one ML) GB segregation. In the zirconia obtained by the conventional sintering methods, a minor amount of silicon, originated from contaminated starting materials, detrimentally influences the conductivity of fuel cells oxides [36]. This effect originates from silicon coverage of GBs in stabilized zirconia with the formation of a continuous GB network in the polycrystal. Silicon-containing phase forms lenticular GB particles and they do not wet the GBs. However, if the Si concentration in GBs reaches about 0.5 ML, the GB conductivity drastically decreases, and does not change much with a further increase of GB Si content [36]. However, if the grain size in stabilized zirconia decreases from micrometer into the nanometer range, the amount of silicon will not be enough to contaminate all GBs. As a result, the specific GB conductivities in nanocrystalline calcia-stabilized zirconia increase about five times [36]. The specific GB conductivities of the nanocrystalline YSZ samples (grain size 40 nm) are 1–2 orders of magnitude higher than those of the microcrystalline samples (grain size 400–1,000 nm) (Fig. 8.5) [37]. Therefore, the detrimental effect of Si-contamination

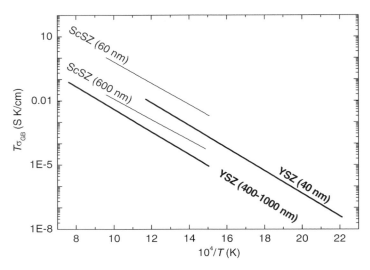

Fig. 8.5 Temperature dependence of specific GB conductivity in nano- and microcrystalline zirconia stabilized by yttria and scandia. Specific GB conductivity increases by decreasing of grain size. Thick lines represent the data of [37] for yttria-stabilized zirconia. Thin lines represent the data of [38] for scandia-stabilized zirconia

vanishes and overall properties of nanostructured zirconia improve. Similar effect of grain size was observed in the scandia-stabilized zirconia [38]. The specific GB conductivities measured using the impedance spectroscopy increase almost two orders of magnitude when grain size decreases from 6,000 to 60 nm (Fig. 8.5). It is an important example, how the GB engineering (tailoring the polycrystal properties by controlling the GB structure and composition) can improve the properties of nanostructured oxides for fuel cells. Thus, by decreasing the grain size, one can dilute the detrimental GB segregation down to the harmless value and improve the properties of a conducting oxide.

8.4.3 Scavengers for GB Impurities

Another way to compensate for the detrimental Si influence and to improve the GB conductivity in zirconia and ceria is to use the so-called scavengers. It has been shown already in 1982 that small additions of Al_2O_3 drastically improve the ionic conductivity of YSZ [39]. Later, Al_2O_3 was identified as the most effective dopant in increasing the GB conductivity of zirconia-based electrolytes [40–43]. Butler and Drennan suggested that alumina acts as a "scavenger" for SiO_2, since the affinity of SiO_2 to Al_2O_3 is greater than to the ZrO_2 [39]. As a result, the particles of Al_2O_3 present in the ceramic "sweep out" silicon from zirconia GBs. It results in the purification effect similar to that of the decrease of grain size. The best scavenger for ceria-based electrolyte is the iron oxide [44].

8.4.4 Heavy Doping

Heavy doping is another way to change the GB composition, and therefore, to improve the conductivity of an oxide. Cerium oxide is a mixed ionic/electronic conductor and exhibits high ionic conductivity when doped with lower valent cations (acceptors). As the oxygen vacancy mobility is even higher than in cubic zirconia – the other prominent fluorite-structured oxygen ion conductor – there has been considerable interest in the potential of ceria-based solid electrolytes for applications in SOFC or oxygen membranes. In [45] the microcrystalline ceria was doped with Y, La, and Gd in the broad concentration range between 0.1 and 27 at.%. The GB effect, which is indicated by the gap between the bulk and the total conductivity, was found to decrease rapidly as the acceptor concentration increases. The GB conductivity drastically increases at the acceptor concentration between 2 and 10 at.% (Fig. 8.6). Simple estimation reveals that the GB conductivity reaches the bulk value when all GBs become covered with a ML of an acceptor impurity (for the ceria grain size of about 1 μm).

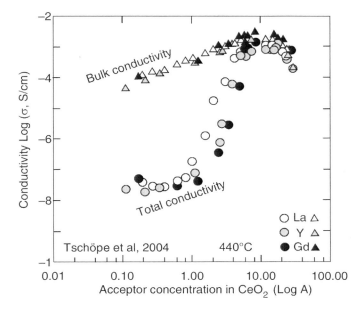

Fig. 8.6 Bulk (*triangles*) and total (*circles*) electrical conductivities at $T = 440°C$ of La-, Y- and Gd-doped microcrystalline cerium oxide as function of dopant concentration according to the data of [45]

8.5 GB Phenomena in Perovskites

Perovskite-type oxides ($BaTiO_3$, $SrTiO_3$, $LaAlO_3$, $LaCrO_3$, etc.) have recently attracted considerable attention for their applications in high-temperature electrochemical devices, such as electrolytes and electrodes of SOFC, oxygen permeating membranes, sensors etc. For the ionic conduction, some perovskites exhibit surprisingly high ionic conductivities, higher than those of well-known zirconia-based materials. The impedance spectroscopy permits to separate bulk and GB inputs in overall conductivity. In many cases, the overall conductivity of perovskites is determined by the GB resistance, like that of Sr and Mg doped $LaAlO_3$ below 550°C [6]. However, the GB input into overall conductivity gradually decreases by increasing temperature. GBs in perovskites mainly contain the conventional GB segregation layer. Only in few cases (like in $BaTiO_3$ sintered from powder particles with Mn coating), the GB amorphous region with a width of about 1 nm was observed [5]. The boundary width in such polycrystals is about five times larger than that in the $BaTiO_3$ sintered from powder particles without Mn coating. The electrostatic potential barrier height of the $BaTiO_3$ ceramics increased from 0.18 to 0.24 eV, due to the increase in the width of the excess negative charge layer from 70 to 120 nm, with an increase in the amount of the powder coating material from 0 to 1.0 at.%. A systematic variation of the GB features with the amount of coating material indicates the possibility of using this synthesis method to get fine control over the chemistry and electrical properties of the semiconducting $BaTiO_3$ ceramic.

8.6 Influence of Synthesis Route on the Properties of Nanostructured Materials

The unique properties of nanostructured materials (including those of nanograined conducting oxides) are of great importance for various advanced applications. However, there are some indications that physical properties of the same material with the same grain size in a nanometer range depends drastically on the preparation technique.

It is well known that during the manufacture of nanostructured materials, the amorphisation may happen, the supersaturated solid solutions may appear, the metastable phases may form [46]. However, there are indications that physical properties of the same material with the same grain size in a nanometer range depend on the preparation technique. The most reliable data on the formation of metastable phases came from the ball-milling experiments. Particularly, the ball milling of steels reliably and reproducibly leads to the dissolution of cementite or the formation of amorphous solid solution in steels [47–50]. Implantation of carbon ions into iron also produces the strongly nonequilibrium structure in surface layers of samples [51]. In other words, ball milling also called mechanical alloying can be compared with a kind of mechanic implantation of one material into another. The high-pressure torsion (HPT, also called compression shear) or deep drawing is principally different from the ball milling. The investigations on HPT of Al-based alloys [52, 53] demonstrated that HPT or deep drawing lead simultaneously (a) to the formation of highly nonequilibrium nanometer grain structure and (b) to the disapperance of nonequlibrium phases and the formation of phases that are in equlibrium at the HPT temperature and pressure. The carefull experiments and analysis of previous publications on HPT demonstrate that HPT leads to the grain refinement, but cannot lead to the disappearance of equlibrium phases or the formation of nonequlibrium phases. It is the most important difference between HPT and ball milling as two technologies for the manufacture of nanostructured materials. Therefore, the application of various novel techniques for the manufacture of nanograined conducting oxides is very promising, especially when they permit to synthesize the novel stable GB phases.

8.7 Synthesis of Nanostructured Oxides by a "Liquid Ceramics" Method

Nowadays the majority of conducting oxides are produced by the sintering of oxide powders. The addition of oxides with low melting points as sintering adds is used for liquid-phase sintering. Sintering has several disadvatages, particularly it includes the high-temperature synthesis steps and leads to the easy contamination of sintered oxides (especially by silicon). New synthesis technologies would permit to broaden the spectrum of oxides and to produce compounds with properties very promising for the SOFCs and electronic components.

Recently, the novel technology has been developed for the deposition of multicomponent oxide films from organic precursors (so-called liquid ceramics) [54]. The films can be deposited on various substrates. The deposited films of ZnO, Y_2O_3, and Ce–Gd–Ni complex oxide are dense, nonporous, nanostructured, uniform, nontextured (Figs. 8.7 and 8.8). Grain size in these films can be varied from

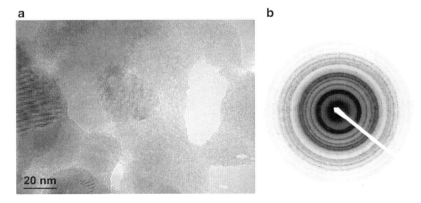

Fig. 8.7 (a) Bright field high-resolution electron micrograph of the nanograined ZnO thin film deposited by the liquid ceramics technology and (b) Electron diffraction pattern. No texture is visible

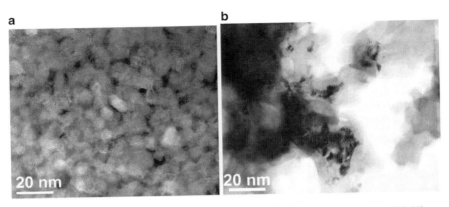

Fig. 8.8 (a) Bright field high-resolution electron micrograph of the nanograined Ce–Gd–Ni complex oxide thin film and (b) of the nanograined Y_2O_3 thin film deposited by the liquid ceramics technology

5 to 100 nm. The components in multicomponent films are distributed uniformly. This technology is extremely flexible. It permits to synthesize the oxides with various compositions and also to change the composition of oxides in the very broad interval. The possibility of tailoring the oxide doping allows one to develop the new advanced materials for the fuel cells and to reach the previously unattainable parameters of the fuel cells. Liquid ceramics method permits to change the grain size and influence the shape of grains (for example, from equiaxial, Fig. 8.8a, to pancake-like, Fig. 8.8b).

8.8 Conclusions

Nanostructured conducting oxides are very promising for various electronic and energy consumption applications like varistors, electrolytes for the SOFC, semipermeable membranes, and sensors. GB phases crucially determine the properties of nanograined oxides produced by powder sintering. GB phase transformations (wetting, prewetting, pseudopartial wetting etc.) proceed in the conducting oxides during sintering and following thermal treatments. Novel GB lines appearing in the conventional bulk phase diagrams permit the GB engineering and tailoring the properties of nanograined conducting oxides. Particularly useful are the novel synthesis methods for conducting oxides, like that of liquid ceramics.

Acknowledgements The authors thank the Russian Foundation for Basic Research (contracts 09-08-90406 and 09-03-92481). They also greatly appreciate Dr. W. Sigle and Dr. F. Phillipp (Max-Planck-Institut für Metallforschung, Stuttgart, Germany) for their help in the electron microscopy investigations.

References

1. H. Wang, Y.-M. Chiang, J. Amer. Ceram. Soc. **81**, 89 (1998)
2. J. Luo, H. Wang, Y.-M. Chiang, J. Amer. Ceram. Soc. **82**, 916 (1999)
3. Y.-M. Chiang, L.A. Silverman, R.H. French, R.M. Cannon, J. Amer. Ceram. Soc. **77**, 1143 (1994)
4. H. Duncan, A. Lasia, Solid State Ionics **176**, 1429 (2005)
5. M.-B. Park, N.-H. Cho, Solid State Ionics **154–155**, 407 (2002)
6. J.Y. Park, G.M. Choi, Solid State Ionics **154–155**, 535 (2002)
7. L.-S. Chang, E. Rabkin, B.B. Straumal, B. Baretzky, W. Gust, Acta mater. **47**, 4041 (1999)
8. B. Straumal, E. Rabkin, W. Lojkowski, W. Gust, L.S. Shvindlerman, Acta mater. **45**, 1931 (1997)
9. S.V. Divinski, M. Lohmann, Chr. Herzig, B. Straumal, B. Baretzky, W. Gust, Phys. Rev. B. **71**, 1041041 (2005)
10. J. Schölhammer, B. Baretzky, W. Gust, E. Mittemeijer, B. Straumal, Interface Sci. **9**, 43 (2001)
11. M. Matsuoka, Jap. J. Appl. Phys. **10**, 736 (1971)
12. L.M. Levinson, H.R. Philipp, Amer. Ceram. Soc. Bull. **65**, 639 (1986)
13. T.K. Gupta, J. Amer. Ceram. Soc. **73**, 1817 (1990)
14. F. Greuter, G. Blatter, Semicond. Sci. Technol. **5**, 111 (1990)
15. B. Bhushan, S.C. Kashyap, K.L. Chopra, J. Appl. Phys. **52**, 2932 (1981)
16. J. Wong, J. Appl. Phys. **51**, 4453 (1980)
17. J. Luo, Y.-M. Chiang, R.M. Cannon, Langmuir **21**, 7358 (2005)
18. J. Wong, J. Amer. Ceram. Soc. **57**, 357 (1974)
19. J. Wong, W.G. Morris, Amer. Ceram. Soc. Bull. **53**, 816 (1974)
20. F. Greuter, Solid State Ionics **75**, 67 (1995)
21. J.P. Gambino, W.D. Kingery, G.E. Pike, H.R. Philipp, J. Amer. Ceram. Soc. **72**, 642 (1989)
22. W.D. Kingery, J.B. van der Sande, T. Mitamura, J. Amer. Ceram. Soc. **62**, 221 (1979)
23. E. Olsson, L.K.L. Falk, G.L. Dunlop, J. Mater. Sci. **20**, 4091 (1985)
24. E. Olsson, G.L. Dunlop, J. Appl. Phys. **66**, 3666 (1989)
25. J.-R. Lee, Y.-M. Chiang, G. Ceder, Acta mater. **45**, 1247 (1997)
26. M. Safronov, V.N. Batog, T.V. Stepanyuk, P.M. Fedorov, Russ. J. Inorg. Chem. **16**, 460 (1971)
27. J.P. Guha, S. Kunej, D. Suvorov, J. Mater. Sci. **39**, 911 (2004)
28. D.R. Clarke, J. Appl. Phys. **49**, 2407 (1978)
29. D.R. Clarke, J. Amer. Ceram. Soc. **70**, 15 (1987)
30. F. Brochard-Wyart, J.-M. di Meglio, D. Quéré, P.G. de Gennes, Langmuir **7**, 335 (1991)
31. H.L. Tuller, A.S. Nowick, J. Electrochem. Soc. **122**, 255–259 (1975)
32. M. Filal, C. Petot, M. Mokchah, C. Chateau, J.L. Charpentier, Solid State Ionics. **80**, 27–35 (1995)
33. G. Petot-Ervas, C. Petot, Solid State Ionics **117**, 27 (1999)
34. A. Rizea, D. Chirlesan, C. Petot, G. Petot-Ervas, Solid State Ionics **146**, 341 (2002)
35. S.P.S. Badwal, Solid State Ionics **76**, 67 (1995)
36. M. Aoki, Y. Chiang, I. Kosacki, L.J. Lee, H. Tuller, Y. Liu, J. Amer. Ceram. Soc. **79**, 1169 (1996)
37. P. Mondal, A. Klein, W. Jaegermann, H. Hahn, Solid State Ionics **118**, 331 (1999)
38. G. Xu, Y.W. Zhang, C.S. Liao, C.H. Yan, Solid State Ionics **166**, 391 (2004)
39. E.P. Butler, J. Drennan, J. Amer. Ceram. Soc. **65**, 474 (1982)
40. M. Godickemier, B. Michel, A. Orlinka, P. Bohac, K. Sasaki, L. Gauckler, H. Henrich, P. Schwander, G. Kostorz, H. Hofmann, O. Frei, J. Mater. Res. **9**, 1228 (1994)
41. A.J. Feighery, J.T.S. Irvine, Solid State Ionics **121**, 209 (1999)
42. A. Yuzaki, A. Kishimoto, Solid State Ionics **116**, 47 (1999)
43. X. Guo, C.Q. Tang, R.Z. Yuan, J. Eur. Ceram. Soc. **15**, 25 (1995)
44. T.S. Zhang, J. Ma, L.B. Kong, S.H. Chan, P. Hing, J.A. Kilner, Solid State Ionics **167**, 203 (2004)
45. A. Tschöpe, S. Kilassonia, R. Birringer, Solid State Ionics **173**, 57 (2004)

46. A.R. Yavari, P.J. Desré, T. Benameur, Phys. Rev. Lett. **68**, 2235 (1992)
47. Y. Xu, M. Umemoto, K. Tsuchiya, Mater. Trans. **9**, 2205 (2002)
48. S. Ohsaki, K. Hono, H. Hidaka, S. Takaki, Scripta Mater. **52**, 271 (2005)
49. G.M. Wang, S.J. Campbell, A. Calka, W.F. Caczmarek, NanoStruc. Mater. **6**, 389 (1995)
50. S.J. Campbell, G.M. Wang, A. Calka, W.F. Caczmarek, Mater. Sci. Eng. A. **226–228**, 75 (1997)
51. S.M.M. Ramos, L. Amarai, M. Behar, G. Marest, A. Vasques, F.C. Zawislak, Radiat. Eff. Def. Sol. **110**, 355 (1989)
52. B.B. Straumal, B. Baretzky, A.A. Mazilkin, F. Phillipp, O.A. Kogtenkova, M.N. Volkov, R.Z. Valiev, Acta Mater. **52**, 4469 (2004)
53. A.A. Mazilkin, B.B. Straumal, E. Rabkin, B. Baretzky, S. Enders, S.G. Protasova, O.A. Kogtenkova, R. Z. Valiev Acta mater. **54**, 3933 (2006)
54. A.A. Myatiev, N.I. Diachenko, A.L. Pomadchik, P.B. Straumal, Nano- and Microsystem Technol. **3**, 19 (2005). In Russian

Chapter 9
Copper Electrodeposition from Ultrathin Layer of Electrolyte

S. Zhong, T. Koch, M. Wang, M. Zhang, and T. Schimmel

Abstract Electrochemical metallization of copper is used in microelectronics e.g., on-chip interconnection. The need for the fundamental understanding of the copper electrodeposition and the avoiding of ramified deposits, by exploration of especially the influence of the chemical–physical environment on the deposition, becomes more essential for the development of the microelectronic- and other related industries. Contrary to standard electrochemistry the electrodeposition from an ultrathin layer of electrolyte produces much more regular deposits, because of suppressed convection noises within this ultrathin layer. The details of the deposit morphology give information about how the copper deposition develops. It also hints at the influence of the local electric field and the local concentration filed on the morphology and the structure of copper deposits. This unique growth system may have significant implications on the pattern formation of many interfacial growth systems. Besides the use for basic scientific research, there are also perspectives for different applications in the field of parallel micro- and nano-wiring and the creation of periodical nanostructured films.

9.1 Introduction

Electrochemical metallization of copper is used in microelectronics as on-chip interconnection due to the copper superior conductivity and high electromigration resistance [1, 2]. However, unfortunately the electrodeposition of metals, such as copper, usually produces ramified deposits [3]. Although even ramified deposits can be useful in some cases [4], the formation of homogenous and uniform deposits is much more important to industrial application [3]. Therefore, there is a great need for a fundamental understanding of the copper electrodeposition. Especially the understanding of the influence of the chemical–physical environment on the deposition process becomes more and more essential. The studies of crystal growth suggest that ramified growth is related to the noises of the external Laplacian fields. It is intriguing to know whether the ramified features remain when the external agitations are suppressed, or regular patterns can be electrodeposited directly. As an example, Monte Carlo simulation shows that by changing the strength of the electromigration, the deposit morphology may vary from fractal to stringy patterns [5]. An experimentally similar tendency was observed in a strong electric field [6,7], where the deposit branches look straight on macroscopic scale, yet on microscopic scale the branches are still ramified, most likely due to the diffusion instabilities. Other experiments show that suppressing the different noises, e.g., by introducing agarose gels, also helps to obtain a more regular morphology on macroscopic scale [8]. However, also in these cases the structures still remain irregular on microscopic scale and in addition with these additives some uncontrollable factors are introduced that make the situation even more complicated. On the other hand, the electrodeposition is also influenced by the interfacial process, which includes cationic electrons and then adsorbing on the interface as metal adatoms. The copper adatoms are then diffusing on the surface of the electrode and finally nucleate into the metal solid phase. This process is sensitive to the variation of the concentration fields and the electric fields at the very close region in the front of interface. Because of the competition of nutrient transport and interfacial kinetics, the interfacial concentration field may become unstable [9–12]. Hence the nucleation rate of the metal fluctuates, and the electrodeposition is modulated by this process.

To optimize the suppression of noises and to explore the basic mechanisms of copper electrodeposition, a unique experimental system using an ultrathin electrolyte layer for electrodeposition has been developed recently [13–19]. In this system, the noises are strongly suppressed and the morphology of deposits changes tremendously. The unique growth behavior in such a system may have significant implications on knowledge about pattern formation in crystal growth systems. The aim of this paper is to discuss recent experimental developments and findings using thin layer electrodeposition and to present potential applications of the unique deposits.

9.2 Experimental Methods

The experimental procedure is summarized in Fig. 9.1. The cell for electrodeposition consists of two parallel glass slides separated by spacers The thin electrolyte is maintained between the two glass slides. Copper wires are used as the electrodes. Two kinds of geometric cells are mainly used: one type is a circular cell, with a circular anode and a point cathode in the center of the cell; the other one is a parallel cell with two parallel wire-like electrodes facing each other. (Fig. 9.1a) The electrolyte is a 0.05 M $CuSO_4$ solution. Before applying the external potential, the electrolyte between the glass slides is solidified carefully by decreasing the temperature to a preset value (e.g. $-3°C$). During solidification, $CuSO_4$ is partially expelled from the ice of electrolyte due to the segregation effect [20]. When equilibrium is reached, an ultrathin aqueous layer of concentrated $CuSO_4$ electrolyte is trapped between the ice interface and the glass slides (Fig. 9.1b). The thickness of electrolyte can be easily tuned from tens of nanometers up to hundreds of nanometers, by changing the preset temperature. The electrodeposition will take place in this

9 Copper Electrodeposition from Ultrathin Layer of Electrolyte

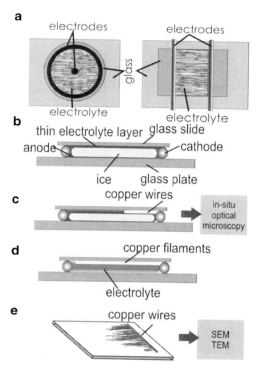

Fig. 9.1 Schematic diagrams showing the process of generating copper deposits. (**a**) Two types of electrodeposition cells. (**b**) Slow and careful freezing of CuSO$_4$ electrolyte. Due to the segregation effect, a thin aqueous layer of electrolyte forms between the upper glass slide and the ice. This makes the concentration of electrolyte in the remaining liquid higher than the initial concentration. (**c**) By applying a constant voltage between the two electrodes, the copper deposits start growing from the cathode into the electrolyte. The process of growth can be observed in situ with optical microscopy. At the end of the experiment, the copper deposits cover the whole electrodeposition cell. (**d**) After deposition, cooling is stopped and the temperature rises. As a result, the ice melts. (**e**) The copper deposits can be taken out of the electrodeposition cell and subsequently be used for further studies and experiments

ultrathin layer of electrolyte. Subsequently a constant voltage or current is applied to the two electrodes. As a consequence, the deposits immerge on the cathode and grow laterally along the surface of the glass substrate into the direction of the anode with a growth rate of several micrometers per second (Fig. 9.1c). When electrodeposition is finished, the temperature increased to melt the ice (Fig. 9.1d). The deposits adhere robustly on the glass substrate and can easily be taken out of the electrodeposition cell for further examinations (Fig. 9.1e). The deposits can be cataloged either to wires or to thin films according to their morphologies.

9.2.1 Copper Submicrowires

Copper wires have been generated using ultrathin layer electrodeposition. The evolution of morphology depending on the varieties of experimental conditions gives the clue to understand the influence of the chemical–physical environment on the Cu-deposition. Much effort has been devoted to the fabrication of micro- and nanostructures with one- or two-dimensional pattern [21–23]. This ultrathin layer technique may pave a new and easy way to the fabrication of such microstructures for basic research and application.

9.2.1.1 Morphology of Wires

The typical copper deposit is shown in Fig. 9.2. Unlike the previously reported random branching morphology [10, 24], here the deposit branches are fingering and have a smooth contour (Fig. 9.2a) [13–15]. The deposit is shiny and grows robustly on the glass substrate. Optical microscopy reveals that the fingering branch consists of "cellular structures" (Fig. 9.2b), and each of them is composed of long, narrow copper wires. Although unbranching long wires (length more than 150 μm) can be found occasionally, bifurcation occurs to most of them. The overall density of the deposit and the average interwire separation do not change evidently, and hence, the cellular pattern gradually increases in width. It can be seen from the tip region that the wires are perpendicular to the contour of the fingering branches, which suggests that the copper wires develop along the local electric field.

Atomic force microscope (AFM) reveals that the finger-like branches actually consist of many straight wires with periodical nanostructures as shown in Fig. 9.3. It is noteworthy that the periodical nanostructures on the neighboring wires correlate

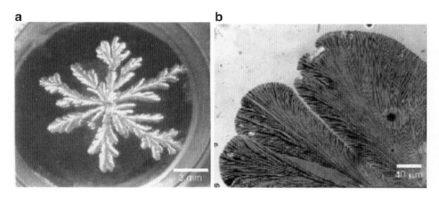

Fig. 9.2 (a) The fingering electrodeposit grown on a glass substrate from an ultrathin film of $CuSO_4$ solution. (b) The optical micrograph of the fingering branches shown in (a), in which fine filaments can be seen. The fingering branches orient to different directions, so the growth of the filaments is not restricted to the specific orientations. The initial concentration of the electrolyte (C) was 0.05 M, the temperature (T) was −4°C, and the applied voltage (V) was 4.0 V. From [13]

9 Copper Electrodeposition from Ultrathin Layer of Electrolyte

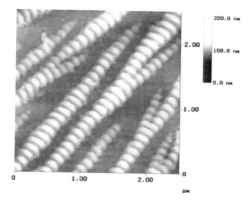

Fig. 9.3 AFM images of the copper filaments (contact mode): The top surface of the filaments is rounded, indicating that it was not confined by any flat, rigid surface during the growth. Periodical structures can be identified on the copper filaments. The periodic corrugations on neighboring filaments are correlated in their position. The parameters for the electro-deposition: $C = 0.05$ M, $V = 4.0$ V, $T = -5°$C. From [13]

in their position, which can be easily identified in the branch-splitting regions. Since the outlines of periodical nanostructures reflect their evolution of growth, the coherence of these structures implies that they were generated simultaneously. Therefore the coherent periodical growth of the wires is associated with an evident oscillation of the electric deposition-current. The periodicity of these spatiotemporal oscillations depends on the voltage between the electrodes, the pH of the electrolyte, the temperature, etc. The distinct difference between the electrodeposits that are shown here and those that were reported previously [14] is that the branching rate has been decreased significantly.

9.2.1.2 Structure and Composition

The structure and chemical composition of the periodical nanostructured wires have been analyzed by transmission electron microscopy (TEM) [13]. Fig 9.4a illustrates the diffraction contrast image of the wires, where the crystallites within each periodical structure are a few tens of nanometers in size. The electron diffraction of the copper wire (Fig. 9.4b) confirms that the microstructure of the wires is polycrystalline. In addition to the diffraction of copper, a diffraction of Cu_2O is also identified. The electric resistivity of the copper wire shows, however, that the average concentration of Cu_2O in the thick part of the wires is less than 2% [15]. We examined the distribution of Cu_2O along a wire by analyzing the diffraction strength at the sites A–F shown in Fig. 9.4a. It turned out that the ratio of the integrated strength of the diffraction of $Cu_2O(111)$ versus that of $Cu(200)$, which is proportional to the ratio of the local content of Cu_2O and Cu, fluctuates as a function of the position (Fig. 9.4c).

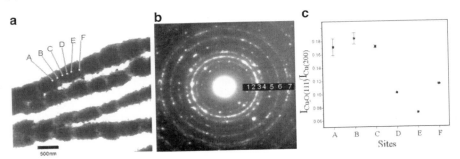

Fig. 9.4 TEM analysis of the periodically structured copper wires. (**a**) The diffraction contrast image of the filaments. (**b**) The electron diffraction of the filament. The numbers on the diffraction rings represent the following materials: (1) Cu_2O (110); (2) Cu_2O (111); (3) Cu (111); (4) Cu (200); (5) Cu_2O (220); (6) Cu (220); (7) Cu (311). (**c**) The ratio of the integration strength of the diffraction of Cu_2O (111) vs. that of Cu (200) at the sites marked by A–F in (**a**). The percentage of Cu_2O is higher at the sites B and F. The parameters for the sample preparation: $C = 0.05$ M, $V = 2.5$ V, $T = -2.75°$C. From [13]

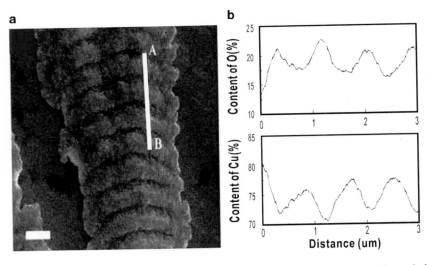

Fig. 9.5 (**a**) SEM image of electrodeposited filaments grown on GaAs substrate. The scale bar represents 1 μm. (**b**) The EDXS-determined content of copper and oxygen elements along the line from the point A to B marked in (**a**). The here measured sample was prepared by applying a voltage of 1 V, the concentration of the electrolyte was 0.05 M. From [16]

A more direct evidence of the concentration of Cu and Cu_2O, fluctuating as a function of position was captured by energy-dispersive X-ray spectrometry (EDXS) and is presented in Fig. 9.5 [16]. The copper wires were deposited on single crystalline GaAs substrate (instead of glass). Figure 9.5a shows the scanning electron microscopy (SEM) image of the periodical nanostructured copper wires. Figure 9.5b illustrates the concentrations of Cu and O along the line marked in Fig. 9.5a.

Corresponding to the periodical structure, the ratio of Cu and O is oscillating. This means that the concentration of Cu_2O in valley regions is higher than that in bump regions and fluctuates during the electrodeposition.

9.2.1.3 Mechanism of Wire Formation

The periodical structures of the copper wires presented here originate from oscillations in electrochemical deposition current. At the moment, we focus on the question whether the ramified feature remains when the external agitations are suppressed. The answer seems to be positive. The crucial difference of the ultrathin layer system and previous ones is that the thickness of the electrolyte is decreased to the order of the length of the mean free path of ion diffusion [13], which leads to the observed distinct morphologies. In the ultrathin film only a small fraction of the cations can move forward without collision with the boundaries. As a matter of fact, the mean free path of ion diffusion (λ) in a thin electrolyte film with two rigid boundaries can be expressed as

$$1/\lambda = 1/\lambda_\infty + 1/L$$

where L is the thickness of the electrolyte film and λ_∞ is the mean free path in a bulk system. If L approaches λ_∞, λ decreases evidently. The diffusion constant D is proportional to λ according to the transport theory. Therefore, ion diffusion in an ultrathin layer is slower. Thus the noise within the system is suppressed significantly and very regular deposits with greatly low branching are generated.

The mass transport depends on the concentration gradient (diffusion) and local electric field (electromigration). Monte Carlo simulation shows that the branches of deposits tend to be much straighter when a strong electric field is applied [5]. It means that the morphology of deposits is determined by cationic diffusion and electromigration. If the electromigration plays a bigger role than the diffusion, the morphology changes from ramified patterns to straight wires. It is important to identify the role of electromigration in an ultrathin layer system. The experiment was performed without an external field by replacing one copper electrode by zinc. The results are shown in Fig. 9.6. The low branching wires also are generated without an external electric field. This implies that the generation of regular deposits is caused by the decrease of the thickness of the electrolyte layer [17].

Even though unbranching long wires (more than 150 μm) can be found occasionally, bifurcation occurs to most of them from the results mentioned earlier. On the other hand the question rises, whether regular patterns can be directly electrodeposited. This question is already important, because much effort has recently been devoted to fabricate micro/nanostructures with regular two-dimensional patterns [21–23]. By carefully controlling the applied voltage, parallel wires can be generated on the silicon substrate at large scale (Fig. 9.7). Two factors are essential for the formation of such parallel wires. One is the local electric field in the vicinity of the tips. The distribution of the electric filed should be uniform and ensure a continuous growth of the tips in a parallel fashion. This can be achieved by careful

Fig. 9.6 The morphology of copper deposits generated without an external electric field. The branching rate is also extremely low and the morphology is similar to that generated by applying an electric field. The results imply that the main factor that determines the morphology is the thickness of electrolyte layer. The parameters for the sample preparation: $C = 0.05\,\text{M}$, $T = -2°\text{C}$. The scale bar in all these images represents $5\,\mu\text{m}$. From [17]

Fig. 9.7 Scanning electron micrographs showing copper wire arrays deposited on a modified silicon oxide surface. The copper wires are straight and uniform in width. The concentration of the $CuSO_4$ electrolyte and the applied voltage were 0.05 M and 1.3 V, respectively. From [18]

control of the applied voltage. When the applied potential reaches the critical value, an optimum electric filed distribution can be achieved. The other factor is the balance between the consumption of cations and the transport of cations. For a proper experiment, it is required that the cations that are moved by the mass transport can slightly overcome the barrier of the depletion layer, which is made of cations due to electrodeposition. This stabilizes the overall physical and chemical conditions in the vicinity of a tip. Otherwise, the first optimal condition will be broken [18].

9.2.2 Periodically Nanostructured Films

Oscillatory growth is commonly observed in electrochemical systems and because of e.g. controllability, attracts the research interest to understand it [25–28]. Using oscillatory systems one can spontaneously generate periodical nanostructures [29]. In previous studies, much attention has been paid to multilayer systems, a periodical structure growing in the direction perpendicular to the substrate. The formation of a periodical structure horizontally in the plane of the substrate has not been studied well to date. Using the similar technique as in Sect. 9.2.1, the periodical nanostructured films could be deposited from an ultrathin layer of electrolyte [19]. These results will support both the understanding of the microscopic processes in electrodeposition and the fabrication of pattern nanostructures by self-organization.

9.2.2.1 Results and Discussion

Figure 9.8 shows the AFM image of a thin periodically nanostructured film. The compact and solid film was deposited on a glass substrate by applying a constant voltage of 1.5 V and setting the temperature to −2.0°C. The perspective image

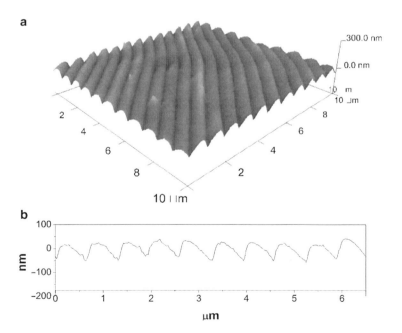

Fig. 9.8 AFM images. (**a**) A perspective image. The copper film shows the wave-like morphology and has a good periodic structure. (**b**) A cross-section of the copper film. The wave-like curve represents the wave-like morphology of the film and clearly illustrates the periodic structure. The modulation wavelength is about 600 nm. The experiment was carried out at $T = -0.3°C$ and $I = 50\,\mu A$

(Fig. 9.8a) shows the wave-like shape of the film. The curve of the cross-section (Fig. 9.8b) illustrates the wave-shape as well. The modulation wavelength in this case is about 600 nm and the average height distance between a peak and a valley is about 62 nm. The modulation wavelength is strongly dependent on the growth conditions (e.g. pH, current, or voltage) and can easily be tuned from 80 nm to a few hundred nanometers [19].

As for copper wires, the formation of the periodical structure is due to the spontaneous alternating growth of Cu and Cu_2O. The crystallites of Cu_2O concentrating in the valley regions of the film are confirmed by the investigation by scanning near-field optical microscopy (SNOM). The light source of the SNOM is an Ar ion laser, which generates three wavelengths: 465, 488, and 514 nm. The light of 514 nm can be strongly absorbed by Cu_2O, whereas it is not absorbed by copper. Therefore this wavelength can be used to distinguish regions of Cu_2O and Cu. Figure 9.9a shows the friction image of the structured film, where the periodical structure can be clearly identified. Figure 9.9b shows the corresponding measured absorption intensity above the same region as it's shown in Fig. 9.9a. The dark bands correspond to the strong absorption of Cu_2O. Obviously the regions with strong absorption in Fig. 9.9b correspond to the valleys of the surface. This result is consistent with results of the copper wire TEM [13].

During the formation of the periodic structures of the film, the voltage between the electrodes is oscillating (galvanostatic mode), as shown in Fig. 9.10a. The Fourier transform of the oscillating voltage is illustrated in the inset, where the primary frequency is 0.2 Hz. The temporal oscillation of the electric signal and the spatial periodicity on the electrodeposits are precisely linked together. The oscillation relates to the applied current across the electrodes and to the pH of

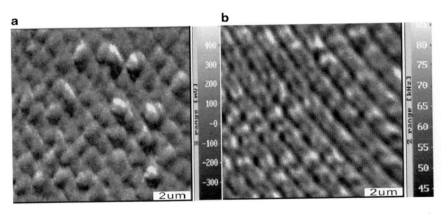

Fig. 9.9 SNOM images of a periodical structured film. (**a**) Friction image providing the morphological information of the film. (**b**) Simultaneous mapping of the absorption intensity over the same area of the film. Comparing (**a**) and (**b**) one may find that the strong absorption at dark strips in (**b**) corresponds to the ditches in the morphology. The wavelength we used in the experiment (514 nm) is within the adsorption band of Cu_2O. We therefore conclude that the crystallites of cuprous oxide are much richer in the valley regions than in the ridge regions. The experimental conditions for the sample preparation: $I = 50\,\mu A$, $T = 0.3°C$, and $C = 0.05\,M$. From [19]

9 Copper Electrodeposition from Ultrathin Layer of Electrolyte   99

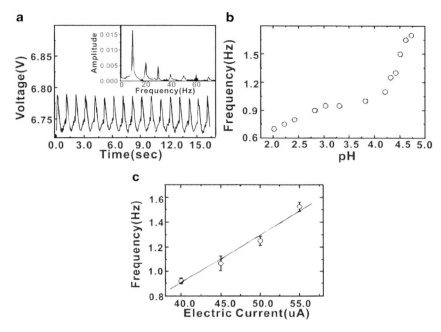

Fig. 9.10 (**a**) The oscillating voltage between the electrodes measured in the galvanostatic mode. The inset is the Fourier transform of the oscillating voltage. The experimental conditions are the following: $I = 50\,\mu A$, $T = -2.0°C$, and $C = 0.05\,M$. (**b**) The dependence of the oscillation frequency as a function of the pH of the electrolyte. The experimental conditions are set as: $I = 40\,\mu A$, $T = -4.7°C$, and $C = 0.05\,M$. From [19]

the electrolyte. As illustrated in Fig. 9.10b, the measured fundamental oscillation frequency decreases when the solution becomes more acidic. If the electrolyte is sufficiently acidic, no oscillations will be observed. The critical pH to ensure the oscillatory growth is 2.0 when the electric current is $40\,\mu A$, the electrolyte concentration is $0.05\,M$, and the temperature is $-2.0°C$. The fundamental oscillation frequency becomes higher when the electric current becomes stronger [19].

9.2.2.2 Mechanism of Oscillation

Electrocrystallization of copper can be understood as follows: Cu^{2+} ions are driven to the cathode by the electric field and the concentration gradient, and then they are reduced and diffuse on the deposit surface as add atoms. Nucleation of the add atoms, followed by limited growth, gives rise to crystallite agglomerates. According to the Nernst equation, the equilibrium potential of $Cu|Cu^{2+}$ increases when the concentration of Cu^{2+} ($[Cu^{2+}]$) builds up. The deposition of copper could take place only when the potential of the cathode is lower than this equilibrium value. The difference between the actual potential and the equilibrium potential (overpotential) is the drive force to the formation of a new phase (Cu^{2+} changes to solid phase directly). Since Cu^{2+} can be reduced to Cu_2O as well, the following two reactions compete with each other.

$$Cu^{2+} + 2e = Cu \tag{9.1}$$
$$Cu^{2+} + OH^- + e = Cu_2O + H_2O \tag{9.2}$$

The standard deposition potential for Cu_2O is much higher than that for Cu. Therefore, for a wide range of the electrolyte concentration, Cu_2O deposits with priority. Suppose that $[Cu^{2+}]$ is initially high at the growing interface, the equilibrium potential for copper deposition will also be high. By applying a sufficiently low potential, both Cu and Cu_2O are deposited. It should be noted that the deposition rate of Cu_2O is proportional to the product of both $[Cu^{2+}]$ and $[OH^-]$, whereas $[OH^-]$ is much lower than $[Cu^{2+}]$. Therefore, the deposition rate of Cu_2O is very low compared with that of copper. The electrodeposition consumes Cu^{2+}. At the same time, the ion transport is confined by the ultrathin electrodeposition system. As a result, $[Cu^{2+}]$ decreases in front of the growing interface, and it takes time for the Laplacian fields to compensate for this reduction. Meanwhile, the equilibrium potential of Cu decreases and it may even become lower than the actual electrode potential. Once this occurs, copper deposition stops, although the deposition of Cu_2O remains. A peak of voltage emerges to maintain the current. At the same time, more Cu^{2+} is driven to the cathode, and as a result, $[Cu^{2+}]$ is accumulated again. Consequently the equilibrium electrode potential of $Cu^{2+}|Cu$ resumes. If its value exceeds the actual electrode potential, copper deposition restarts and the voltage drops down to keep the current constant. In this way, the copper film with periodically modulated concentration of Cu_2O and hence periodical nanostructures are generated [13, 18, 19].

9.3 Conclusion

The ultrathin layer electrodeposition makes a significant improvement toward the understanding of ramified phenomena in metal electrochemical deposition. Promising perspectives of such systems in the domain of nanowires and nanostructured films have been demonstrated here.

References

1. D. Edelstein et al., Tech. Dig. Int. Electron. Devices Meet. 773 (1997)
2. S. Venkatesan et al., Tech. Dig. Int. Electron. Devices Meet. 769 (1997)
3. J.W. Dini, *Electrodeposition: The Materials Science of Coatings and Substrates* (Noyes, LLC, New York, 1993)

4. H.C. Shin, J. Dong, M.L. Liu, Adv. Mater. **15**, 1610 (2003)
5. J. Erlebacher, P.C. Searson, K. Sieradzki, Phys. Rev. Lett. **71**, 3311 (1993)
6. J.R. Melrose, D.B. Hibbert, R.C. Ball, Phys. Rev. Lett. **65**, 3009 (1990)
7. M.Q. Lopez-Salvans, F. Sagues, J. Claret, J. Bassas, Phys. Rev. E **56**, 6869 (1997)
8. M. Wang, N.B. Ming, P. Bennema, Phys. Rev. E **48**, 3825 (1993)
9. V. Fleury, Nature (London) **390**, 145 (1997)
10. V. Fleury, D. Barkey, Europhys. Lett. **36**, 253 (1996)
11. S.N. Atchison, R.P. Burford, D.B. Hibbert, J. Electroanal. Chem. **371**, 137 (1994)
12. M. Wang et al., Phys. Rev. E **60**, 1901 (1999)
13. M. Wang, S. Zhong, X.B. Yin, J.M. Zhu, R.W. Peng, Y. Wang, K.Q. Zhang, N.B. Ming, Phys. Rev. Lett. **86**, 3827 (2001)
14. S. Zhong, Y. Wang, M. Wang, M.Z. Zhang, X.B. Yin, R.W. Peng, N.B. Ming, Phys. Rev. E **67**, 061601 (2003)
15. S. Zhong, M. Wang, X.B. Yin, J.M. Zhu, R.W. Peng, Y. Wang, N.B. Ming, J. Phys. Soc. Jpn. **70**, 1452 (2001)
16. Z. Wu et al., J. Phys.: Condens. Matter **18**, 5425 (2006)
17. Y.Y. Weng, J.W. Si, W.T. Gao, Z. Wu, M. Wang, R.W. Peng, N.B. Ming, Phys. Rev. E **73**, 051601 (2006)
18. M.Z. Zhang, G.H. Zuo, Z.C. Zong, H.Y. Cheng, Z. He, C.M. Yang, G.T. Zou, Small **2**, 727 (2006)
19. Y. Wang, Y. Cao, M. Wang, S. Zhong, M.Z. Zhang, Y. Feng, R.W. Peng, X.P. Hao, N.B. Ming, Phys. Rev. E **69**, 021607 (2006)
20. F. Rosenberger, *Fundamentals of Crystal Growth* (Springer, Berlin, 1979)
21. J. Elezgaray, C. Leger, F. Argoul, Phys. Rev. Lett. **84**, 3129 (2000)
22. I. Mukhopadhyay, W. Freyland, Langmuir **19**, 1951 (2003)
23. G.M. Whitesides, Small **1**, 172 (2005)
24. T. Vicsek, *Fractal Growth Phenomena*, 2nd edn (World Scientific, Singapore, 1992), and references therein
25. J. Wotjowicz, in *Modern Aspects of Electrochemistry*, vol. **8**, ed. by J.O.M. Bockris B.E. Conway (Plenum, New York, 1972), p. 47
26. F.N. Albahadily M. Shell, J. Chem. Phys. **88**, 4312 (1988)
27. F. Argoul A. Kuhn, J. Electroanal. Chem. **359**, 81 (1993)
28. R.M. Suter P.Z. Wong, Phys. Rev. B **39**, 4536 (1989)
29. J.A. Switzer, C.J. Hung, L.Y. Huang, E.R. Switzer, D.R. Kammler, T.D. Golden, E.W. Bohannan, J. Am. Chem. Soc. **120**, 3530 (1998)

Chapter 10
Effect of Plasma Environment on Synthesis of Vertically Aligned Carbon Nanofibers in Plasma-Enhanced Chemical Vapor Deposition

Igor Denysenko, Kostya Ostrikov, Nikolay A. Azarenkov, and Ming Y. Yu

Abstract We present a theoretical model describing a plasma-assisted growth of carbon nanofibers (CNFs), which involves two competing channels of carbon incorporation into stacked graphene sheets: via surface diffusion and through the bulk of the catalyst particle (on the top of the nanofiber), accounting for a range of ion- and radical-assisted processes on the catalyst surface. Using this model, it is found that at low surface temperatures, T_s, the CNF growth is indeed controlled by surface diffusion, thus quantifying the semiempirical conclusions of earlier experiments. On the other hand, both the surface and bulk diffusion channels provide a comparable supply of carbon atoms to the stacked graphene sheets at elevated synthesis temperatures. It is also shown that at low T_s, insufficient for effective catalytic precursor decomposition, the plasma ions play a key role in the production of carbon atoms on the catalyst surface. The model is used to compute the growth rates for the two extreme cases of thermal and plasma-enhanced chemical vapor deposition of CNFs. More importantly, these results quantify and explain a number of observations and semiempirical conclusions of earlier experiments.

10.1 Introduction

Vertically aligned carbon nanotubes (CNTs) and CNFs have high aspect ratios, are mechanically and chemically robust conductors of electrons, and can be deterministically produced on any substrate [1, 2]. They are utilized as supercapacitors, as filler in composite materials, as catalyst support, for hydrogen storage, as electrodes for fuel cells, in field-emission devices, or for ultrafiltration membranes [1–3].

The carbon nanostructures are commonly synthesized using catalyzed chemical vapor deposition (CVD), arc discharge based and radiofrequency magnetron sputtering, etc. Using plasma-enhanced CVD (PECVD), it turns out to be possible to grow CNFs with a better vertical alignment in addition to the improved size and spatial uniformity at higher deposition rates and substrate temperatures remarkably lower than that in most neutral gas-based processes. It is commonly accepted that processes on the surface and within the metal catalyst nanoparticle on top of a CNF determine the subsequent growth and structure. However, how exactly the plasma environment (e.g., ion bombardment and plasma etching [4,5]) affects these processes and translates into (1) higher growth rates, (2) lower activation energies for CNF growth, and (3) lower growth temperatures, remains essentially unclear despite extensive efforts to explain the growth kinetics or to invoke the modeling of neutral gas-based CVD, atomistic structure of related nanoassemblies, or a limited number of plasma-related effects (ion/radical composition, surface heating, etc.). This issue remains one of the major obstacles for deterministic plasma-aided synthesis of CNFs and related nanostructures.

In this paper, we report a model of plasma-assisted growth of CNFs, which involves two competing channels of carbon incorporation into stacked graphene sheets: via diffusion on the surface and through the bulk of the catalyst particle. This model accounts for a range of ion- and radical-assisted processes on the catalyst surface that are unique to plasma environments yet are sidestepped by the existing models of CNF/CNT growth [6–9]. It is found that at low surface temperatures, T_s, the CNF growth is indeed controlled by the surface diffusion, which quantifies the semiempirical conclusion of earlier experiments [10,11]. On the other hand, both the surface and bulk diffusion channels provide a comparable supply of carbon atoms to the stacked graphene sheets at elevated synthesis temperatures. It is also shown that at low T_s, insufficient for effective catalytic precursor decomposition, the plasma ions play a key role in the production of carbon atoms on the catalyst surface. Here, we also quantify the effect of the ion bombardment of the catalyst surface and relate it to a remarkably lower CNF growth activation energy in the plasma-based process, which has been a highly debated yet still has been an intractable issue. The model enabled us to elucidate the predominant channels of carbon incorporation into the stacked graphene sheets and to compute the deposition rates in the two extreme cases of thermal and plasma-enhanced CVD of CNFs. More importantly, these results quantify and explain a number of observations and semiempirical conclusions of earlier experiments [10–13].

10.2 Theoretical Model

Let us consider the plasma-assisted grown of a CNF with a metal catalyst particle on top as shown in Fig. 10.1.

It is assumed that carbon atoms, the primary building units of the nanofibers, are created on the flat and circular top surface of the particle via a number of elementary processes (as sketched in Fig. 10.1) and then incorporated into the growing graphene sheets (shown as stacked cones in Fig. 10.1) via surface or bulk diffusion. By doing so, we calculate the CNF growth rate, H_t, single out specific contributions of the two competing diffusion processes (H_s and H_υ for the surface and bulk diffusion, respectively), and apply these rates to explain and quantify the relevant experimental results [10–13].

The total CNF growth rate $H_t = H_s + H_\upsilon$ can be split into two components originating from the surface $H_s = m_C J_s / \left(\pi r_p^2 \rho \right)$ and bulk $H_\upsilon = m_C J_\upsilon / \left(\pi r_p^2 \rho \right)$

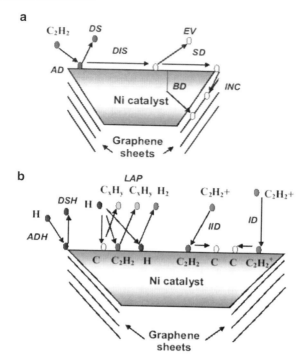

Fig. 10.1 Processes that are common for thermal CVD and PECVD (**a**) and the additional processes on the catalyst surface that are accounted for in the PECVD (**b**). AD = adsorption of C_2H_2; DS = desorption of C_2H_2 (activation energy E_{aCH}); DIS = dissociation (δE_i); EV = evaporation (E_{ev}); SD = surface diffusion (E_s); INC = incorporation into a graphene sheet (δE_{inc}); BD = bulk diffusion (E_b); ADH = adsorption of H; DSH = desorption of H (activation energy E_{aH}); LAP = loss of adsorbed particles at interaction with atomic hydrogen; IID = ion-induced dissociation of C_2H_2; and ID = $C_2H_2^+$ ion decomposition

diffusion, where J_s and J_v are the fluxes of carbon atoms to the graphene sheets over the catalyst particle's surface and bulk respectively. Here, r_p is the particle's radius, $\rho \approx 2\,\text{g/cm}^3$ is the CNF material density, and m_C is the mass of a carbon atom. The flux of C atoms through the catalyst bulk is $J_v = \int_0^{r_p} \left(n_C D_b / r_p^2\right) 2\pi r\, dr$, where n_C is the surface density of carbon atoms, $D_b = D_{b0}\exp\left(-E_b/k_B T_s\right)$ is the bulk diffusion coefficient with D_{b0} a constant and $E_b \approx 1.6\,\text{eV}$ [10], and k_B is the Boltzmann's constant. To calculate the surface diffusion flux

$$J_s = -D_s \frac{dn_C}{dr}\bigg|_{r=r_p} \times 2\pi r_p,$$

we have assumed that diffusing carbon atoms are incorporated into the graphene sheet at the border of the catalyst particle $(r = r_p)$, with the rate determined from $-D_s\, dn_C/dr = k n_C$, where $D_s = D_{s0}\exp\left(-E_s/k_B T_s\right)$ is the surface diffusion coefficient, D_{s0} is a constant, E_s is the energy barrier for carbon diffusion on the catalyst surface, $k = A_k \exp(-\delta E_{inc}/k_B T_s)$ is the incorporation constant, and A_k is the constant that depends on the carbon nanostructure size [14]. Here, δE_{inc} is the barrier for C diffusion along the graphene–catalyst interface; it is ~ 0.4 eV for the graphene–Ni interface [10].

To calculate the number density of carbon atoms, n_C, we assume that the top surface of the catalyst nanoparticle is subject to incoming fluxes of hydrocarbon neutrals (here, C_2H_2), an etching gas (here, H) and hydrocarbon ions (here, $C_2H_2^+$). The surface coverage by C_2H_2, C, and H species is θ_{CH}, θ_C, and θ_H, respectively. Similar to the established CVD growth models of CNTs and related structures, our model accounts for the adsorption and desorption of C_2H_2 and H as well as thermal dissociation of C_2H_2 on the catalyst surface (Fig. 10.1a). We also account for evaporation of carbon atoms from the catalyst surface [15]. To describe the plasma-based CNF growth, we also included the following processes on the catalyst surfaces, unique to the plasma environments yet not accounted for in the existing models: ion-induced dissociation of C_2H_2, interaction of all the adsorbed species with incoming hydrogen atoms, and dissociation of hydrocarbon ions (Fig. 10.1b). More specifically, the model includes the following mass balance equations

$$J_C + \text{div}(D_s \text{ grad } n_C) - O_C = 0, \quad (10.1)$$

$$Q_{CH} - \theta_{CH} j_i y_d - n_{CH} \nu \exp(-\delta E_i/k_B T_s) = 0, \quad (10.2)$$

and

$$Q_H + 2n_{CH} \nu \exp(-\delta E_i/k_B T_s) = 0, \quad (10.3)$$

for C, C_2H_2, and H species on the catalyst surface, respectively. In (10.1), $J_C = 2n_{CH}\nu \exp(-\delta E_i/k_B T_s) + 2\theta_{CH} j_i y_d + 2j_i$ is the carbon source term describing generation of C on the catalyst due to thermal and ion-induced dissociation of C_2H_2, and decomposition of $C_2H_2^+$, respectively. The second term in (10.1) describes the carbon loss due to surface diffusion. Likewise, $O_C = n_C \nu \exp(-E_{ev}/k_B T_s) + n_C \sigma_{ads} j_H + n_C D_b/r_p^2$ accounts for the carbon atom loss due to evaporation (with the energy barrier E_{ev}), interaction with atomic hydrogen from the plasma, and bulk diffusion. In (10.2) and (10.3), $Q_\alpha = j_\alpha(1-\theta_t) - n_\alpha \nu \exp(-E_{a\alpha}/k_B T_s) - n_\alpha \sigma_{ads} j_H$ and subscript $\alpha = (CH, H)$ stands for C_2H_2 and H species, respectively; $n_\alpha = \theta_\alpha \nu_0$ is the surface concentration of species α; $\nu_0 \approx 1.3 \times 10^{15}$ cm^{-2} [16] is the number of adsorption sites per unit area; and $\nu \approx 10^{13}$ Hz is the thermal vibrational frequency. Furthermore, $E_{a\alpha}$ is the desorption activation energy for species α, $\sigma_{ads} \approx 6.8 \times 10^{-16}$ cm^2 is the cross-section for the reactions of atomic hydrogen with adsorbed particles [16], and $\theta_t = \theta_{CH} + \theta_H + \theta_C$ is the total surface coverage. The first, second, and third terms in the expression for Q_α describe the adsorption, desorption of species α, and interaction of the adsorbed species α with atomic hydrogen (with the incoming flux j_H) from the plasma, respectively. The flux of the impinging species is given by $j_\alpha = \tilde{n}_\alpha \nu_{th\alpha}/4$, where $\tilde{n}_\alpha$ and $\nu_{th\alpha}$ are the volume density and thermal velocity, respectively. The second term in (10.2) accounts for C_2H_2 loss due to ion bombardment, where $j_i \approx n_i \sqrt{k_B T_e/m_i}$ is the ion flux, n_i

is the ion density in the plasma, T_e (≈ 1.5 eV) is the electron temperature, m_i is the ion mass, $y_d \approx 2.49 \times 10^{-2} + 3.29 \times 10^{-2} E_i$ [16], and E_i is the ion energy in electron volts. Here, we have assumed that the C_2H_2 loss due to ion-induced dissociation is the same as that in the growth of diamond-like films [16]. The last terms in (10.2) and (10.3) account for C_2H_2 loss and H generation as a result of thermal dissociation of acetylene with the energy barrier δE_i.

10.3 Results and Discussion

We now present and discuss the numerical results that follow from the aforementioned model. To elucidate the relative roles of the surface and bulk diffusion channels under typical experimental conditions of CNF growth by CVD [12] and PECVD [10, 11], we studied the dependence of H_t, H_s, and H_υ on the surface temperature. Figure 10.2 (a) shows the comparison of the CNF growth rates computed here [using $\tilde{n}_{CH} = 5 \times 10^{14}$ cm^{-3}, $\tilde{n}_H = 10^{-3} \tilde{n}_{CH}$, $E_i = 500$ eV, $n_i = 3 \times 10^{10}$ cm^{-3}, $r_p = 25$ nm, $E_{aCH} = 2.9$ eV, $E_{aH} = 1.8$ eV, $\delta E_i = 1.3$ eV, $E_s = 0.3$ eV, and $\delta E_{inc} = 0.4$ eV] and measured experimentally by Hofmann et al. [11].

It is seen that the calculated H_t successfully reproduces the experimental trend in the CNF growth rate with slightly increasing deviations at low ($\beta_T = 1,000$ K/$T_s > 1.9$) and large substrate temperatures ($\beta_T < 1.1$). The most striking observation is that the surface diffusion curve fits best to the experimental curve in the broad range of temperatures ($\beta_T > 0.9$). The minor difference at low T_s may be attributed to the heating of Ni catalyst particles by intense ion fluxes from the plasma [17]. This confirms and quantifies the earlier conclusion [10] that the CNF synthesis in the experiments of Hofmann et al. [11] may indeed be due to surface diffusion of carbon atoms over the catalyst particle surface.

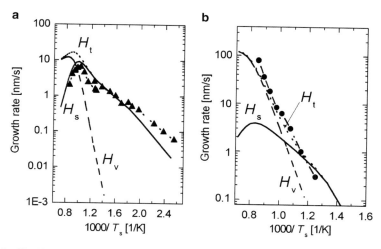

Fig. 10.2 H_s, H_υ, and H_t as functions of the substrate temperature for PECVD (**a**) and CVD (**b**). Triangles and circles correspond to experimental points that are taken from [11, 12]

More importantly, at low substrate temperatures the temperature dependence of the CNF growth rate due to surface diffusion $H_s \sim \exp(-\delta E_{inc}/k_B T_s)$ appears to be the same as that of the constant k of carbon incorporation into graphene sheets. We thus conclude that the activation energy in PECVD is about the barrier for carbon diffusion along the graphene–catalyst interface. Given that δE_{inc} is only ~ 0.4 eV [10], this very low activation energy of the plasma-based growth of CNF in fact explains higher growth rates in plasma-aided process compared with CVD [11].

Let us now consider the CVD case by letting $j_H = j_i = 0$. The dependencies of H_t, H_s, and H_υ on β_T, calculated for $\tilde{n}_{CH} = 10^{16}$ cm^{-3}, $r_p = 30$ nm, $\delta E_{inc} = 0.5$ eV, and the other parameters remaining the same as in Fig. 10.2a, are displayed in Fig. 10.2b. From Fig. 10.2b, one notices that the computed total growth rate, H_t, is indeed very close to the experimental results of Ducati et al. [12]. It is clearly seen that at lower temperatures ($\beta_T > 1.2$), surface diffusion controls the growth ($H_t \approx H_s$), whereas at higher temperatures ($\beta_T < 1.0$), CNF growth is due to the bulk diffusion ($H_t \approx H_\upsilon$). However, in the intermediate range $1.0 < \beta_T < 1.2$, both growth channels make comparable contributions. Moreover, the asymptotic analysis of the total growth rate, H_t, suggests that the activation energy for the CNF growth is approximately equal to the barrier $\delta E_i \approx 1.3$ eV of thermal dissociation of C_2H_2 at relatively low ($T_s < 700$ K) temperatures and is approximately equal to $E_b \approx 1.6$ eV at high ($T_s > 1{,}000$ K) temperatures. Therefore, our calculations rigorously confirm that the energy barrier for the CNF growth in a plasma is indeed several times lower than that in a CVD process.

We have also studied how the ion and atomic hydrogen fluxes from the plasma affect the CNF growth rate H_t. The growth rates, H_t, as functions of T_s are presented in Figs. 10.3a, b for different ion and hydrogen atom densities in the plasma. One can see from Fig. 10.3a that at low substrate temperatures, the growth rate increases

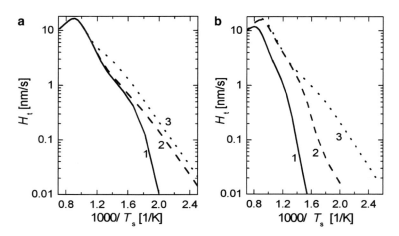

Fig. 10.3 Variation of H_t with T_s for different densities of ions n_i (**a**) and atomic hydrogen $\tilde{n}_H$ (**b**) in the plasma bulk. Curves 1, 2, and 3 in (**a**) correspond to $n_i = 10^8$, 10^{10}, and 10^{11} cm^{-3}, respectively. Curves 1, 2 and 3 in (**b**) are for $\tilde{n}_H = \tilde{n}_{CH}$, $0.05\tilde{n}_{CH}$, $5 \times 10^{-4}\tilde{n}_{CH}$, respectively, where $\tilde{n}_{CH}$ is the density of C_2H_2 species in the plasma bulk. Here, $\tilde{n}_H = 10^{-3}\tilde{n}_{CH}$ for (**a**) and $n_i = 3 \times 10^{10}$ cm^{-3} for (**b**). Other parameters are the same as in Fig. 10.2a

with j_i. This increase is mostly due to the enhanced ion-induced dissociation of C_2H_2 on the catalyst nanoparticles. On the other hand, Fig. 10.3b suggests that H_t decreases with j_H because of the larger loss of C_2H_2 and C species in reactions with impinging hydrogen atoms (Fig. 10.1b).

10.4 Conclusions

Thus, we have developed an advanced model of the plasma-assisted growth of carbon nanofibers. The model accounts for the diffusion of carbon on the catalyst surface and through the metal catalyst particle, as well as carbon generation on the catalyst due to ion bombardment. Using this model, we have found that at low substrate temperatures, the CNF growth is mainly due to the surface diffusion of carbon atoms about the Ni catalyst surface. At relatively large substrate temperatures ($T_s > 800$ K), both surface and bulk diffusion may be important in the CNF growth. At low T_s, the generation and loss of C on the catalyst surface in a plasma-based process is mainly due to ion and etching gas deposition on the catalyst. As a result, in PECVD the dependence of the growth rate on T_s is the same as that of the carbon incorporation constant, which leads to a remarkably low (~ 0.3 eV) activation energy for CNF growth. On the other hand, in the CVD process, the CNF growth heavily relies on the thermal dissociation of hydrocarbons on the catalyst surface and the bulk diffusion of carbon through the catalyst particle. The dissociation and bulk diffusion activation energies are larger than the barrier for C diffusion along the graphene–catalyst interface, and, as a result, the activation energy in PECVD is smaller than in CVD.

Moreover, we have shown that ion-assisted dissociation of hydrocarbon neutrals on the catalyst and decomposition of hydrocarbon ions upon their deposition onto the surface may be the main processes that are responsible for carbon production at low substrate temperatures. This is consistent with the experimental results of Tanemura et al. [13] suggesting that carbon nanofibers do not grow when an ion-repelling positive potential is applied to the substrate.

To conclude, we note that the results and the model of this paper can be used for optimizing carbon nanofiber and nanotube synthesis and we can eventually bring it to the as-yet-elusive deterministic level. More importantly, the main conclusions are not restricted to CNTs or CNFs and may be relevant to the plasma-assisted catalyzed growth of a broader range of nanoassemblies.

Acknowledgments This work was partially supported by the Humboldt Foundation and the Australian Research Council.

References

1. N. Grobert, Mater. Tod. **10**, 28 (2007)
2. A.V. Melechko, V.I. Merkulov, T.E. McKnight, M.A. Guillorn, K.L. Klein, D.H. Lowndes, M.L. Simpson, J. Appl. Phys. **97**, 041301 (2005)
3. S.G. Rao, L. Huang, W. Setyawan, S. Hong, Nature **425**, 36 (2003)
4. M. Meyyappan, L. Delzeit, A. Cassell, D. Hash, Plasma Sources Sci. Technol. **12**, 205 (2003)
5. I.B. Denysenko, S. Xu, J.D. Long, P.P. Rutkevych, N.A. Azarenkov, K. Ostrikov, J. Appl. Phys. **95**, 2713 (2004)
6. J.C. Charlier, A. DeVita, X. Blase, R. Car, Science **275**, 646 (1997)
7. A.N. Andriotis, M. Menon, G. Froudakis, Phys. Rev. Lett. **85**, 3193 (2000)
8. X. Fan, R. Buczko, A.A. Puretzky, D.B. Geohegan, J.Y. Howe, S.T. Pantelides, S. J. Pennycook, Phys. Rev. Lett. **90**, 145501 (2003)
9. C. Klinke, J.-M. Bonard, K. Kern, Phys. Rev. B **71**, 035403 (2005)
10. S. Hofmann, G. Czanyi, A.C. Ferrari, M.C. Payne, J. Robertson, Phys. Rev. Let. **95**, 036101 (2005)
11. S. Hofmann, C. Ducati, J. Robertson, B. Kleinsorge, Appl. Phys. Lett. **83**, 135 (2003)
12. C. Ducati, I. Alexandrou, M. Chowalla, G.A.J. Amaratunga, J. Robertson, J. Appl. Phys. **92**, 3299 (2002)
13. M. Tanemura, K. Iwata, K. Takahashi, Y. Fujimoto, F. Okuyama, H. Sugie, V. Filip, J. Appl. Phys. **90**, 1529 (2001)
14. O.A. Louchev, T. Laude, Y. Sato, H. Kanda, J. Chem. Phys. **118**, 7622 (2003)
15. O.A. Louchev, C. Dussarrat, Y. Sato, J. Appl. Phys. **86**, 1736 (1999)
16. N.V. Mantzaris, E. Gogolides, A.G. Boudouvis, A. Rhallabi, G. Turban, J. Appl. Phys. **79**, 3718 (1996)
17. K.B.K. Teo et al., Nano Lett. **4**, 921 (2004)

Part III
Nanoelectronics

Chapter 11
Single-Atom Transistors: Switching an Electrical Current with Individual Atoms

Christian Obermair, Fangqing Xie, Robert Maul, Wolfgang Wenzel, Gerd Schön, and Thomas Schimmel

Abstract Single-atom transistors are a novel approach opening intriguing perspectives for quantum electronics and logics at room temperature. They are based on the stable and reproducible operation of atomic-scale switches, which allow us to open and close an electrical circuit by the controlled reconfiguration of silver atoms within an atomic-scale junction. We demonstrate the operation of such atomic quantum switches, and discuss in more detail the process during which these switches are formed by repeated electrochemical deposition and dissolution. Only after repeated deposition/dissolution cycles, a bistable contact is formed on the atomic scale, which allows to switch between a configuration where the contact is closed, the conducting state or "on"-state, and a configuration where the contact is open, the nonconducting state or "off"-state. The controlled fabrication of these well-ordered atomic-scale metallic contacts is of great interest: it is expected that the experimentally observed high percentage of point contacts with a conductance at noninteger multiples of the conductance quantum $G_0 = 2e^2/h$ ($\approx 1/12.9\,\mathrm{k}\Omega$) in conventional experiments with simple metals is correlated with defects resulting from the fabrication process. Our combined electrochemical deposition and annealing method allows the controlled fabrication of point contacts with preselectable *integer* quantum conductance. The resulting conductance measurements on silver point contacts are compared with tight-binding-like conductance calculations of modeled idealized junction geometries between two silver crystals with a predefined number of contact atoms.

11.1 Introduction

Due to their interesting physical properties and technological perspectives, atomic-scale metallic point contacts are an object of intensive investigations by numerous groups [1–23]. As the size of these constrictions is smaller than the scattering length of the conduction electrons, transport through such contacts is ballistic and as the width of the contacts is on the length scale of the electron wavelength, the quantum nature of the electron is directly observable. The conductance is quantized in multiples of $2e^2/h$, where e is the charge of an electron and h is Planck's constant. Experimentally, two different approaches are available for the fabrication of these metallic quantum point contacts: mechanically controlled deformation of thin metallic junctions [12–16] and electrochemical fabrication techniques [17–23]. In both the cases, conductance quantization was observed experimentally even at room temperature.

As the contact in such junctions is ultimately formed by only one or a few individual atoms, it has already been suggested that the controlled movement of the contacting atom(s) could lead to a switch or a relay on the atomic scale. By moving only one atom in and out of position, a quantized electrical current could be switched on and off. Such a development would not only mean that the ultimate limit for the size of a switch would be reached, but also would provide the basic functional unit for the potential future field of quantum electronics. However, building such a device will require reproducible control of the position of individual atoms within the quantum point contact by means of a third electrode, the control electrode, or gate electrode.

In the past, first experiments demonstrating controlled switching of atomic positions were reported. Eigler et al. [24] showed the switching of the position of a xenon atom between a scanning tunneling microscope (STM) tip and a nickel sample surface by applying voltage pulses between STM tip and sample. Fuchs and Schimmel [25] demonstrated the switching of atomic positions in a solid surface at room temperatures by applying voltage pulses with the tip of an STM. Another approach was performed recently by Terabe et al. [10], who switched the conductance across a small gap containing a solid electrolyte by voltage pulsing. In these experiments, however, there was no way of controlling the conductance across the atomic-scale junction by means of an independent third electrode, the control electrode or "gate" electrode. Such an independent control electrode, however, is necessary for the fabrication of transistors or relays on the atomic scale.

Recently, we could demonstrate the first implementation of a transistor on the atomic scale [9, 26]. The atomic-scale transistor can be reversibly switched between a quantized conducting on-state and an insulating off-state by applying a control potential relative to a third, independent gate electrode. For this purpose, an atomic-scale point contact is formed by electrochemical deposition of silver within a nanoscale gap between two gold electrodes, which subsequently can be dissolved and redeposited, thus allowing to open and close the gap. However, it turned out that only after numerous such deposition/dissolution cycles, a contact is formed which reproducibly opens and closes as a function of the voltage applied to the gate electrode. Here, we demonstrate the effect of this electrochemical cycling process, and we discuss the mechanisms of formation and operation of the atomic-scale quantum transistor.

11.2 Experimental

The experimental set-up is shown in Fig. 11.1a. By applying an electrochemical potential, silver was deposited within the gap between two macroscopic gold electrodes (gap width typically of the order of 50 nm). The gold electrodes (thickness

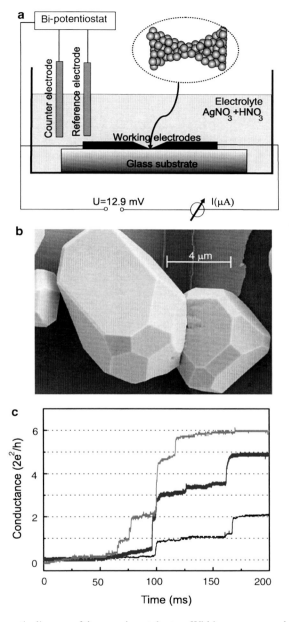

Fig. 11.1 (a) Schematic diagram of the experimental setup. Within a narrow gap between two gold electrodes on a glass substrate, a silver point contact is deposited electrochemically. (b) SEM image of two electrochemically deposited silver crystals between which the atomic-scale silver contact forms (deposition voltage: 30 mV). (c) Conductance-vs.-time curves of three different silver point contacts during initial electrochemical deposition. *Before* electrochemical annealing, contacts of limited stability are formed, typically exhibiting conductance values which are *noninteger* multiples of G_0 (cf. [2])

approx. 100 nm) were covered with an insulting polymer coating except for the immediate contact area and served as electrochemical working electrodes. Two silver wires (0.25 mm in diameter and 99.9985% purity) served as counter and quasi-reference electrodes. The potentials of the working electrodes with respect to the quasi-reference and counter electrodes were set by a computer-controlled bipotentiostat. The electrolyte consisted of 1 mM $AgNO_3$ + 0.1 M HNO_3 in bidistilled water. All experiments were performed at room temperature, the electrolyte being kept in ambient air. For conductance measurements, an additional voltage of 12.9 mV was applied between the two gold electrodes. While one of the gold electrodes was connected to the ground potential, the other gold electrode was kept at -12.9 mV relative to this ground potential.

When applying an electrochemical potential of 10–40 mV between the electrochemical reference electrode and the two gold electrodes (gold electrodes with negative bias relative to the electrochemical reference electrode), silver crystals formed on the two gold electrodes, two crystals finally meeting each other by forming an atomic-scale contact (see Fig. 11.1b). During deposition, the conductance between the two gold electrodes was continuously measured. As soon as a predefined conductance value was exceeded, the computer-controlled feedback immediately stopped further deposition of silver on the working electrodes. If desired, the deposited contact could be fully or partially electrochemically dissolved by applying an electrochemical potential of -15 to -40 mV.

Figure 11.1c gives conductance-vs.-time curves of the closing processes of three different atomic-scale contacts during initial deposition, i.e., before electrochemical annealing. In this way, initially, contacts of limited stability were formed, typically exhibiting conductance values that are *non*integer multiples of G_0.

11.3 Configuring a Bistable Atomic Switch by Repeated Electrochemical Cycling

After the deposition of a silver point contact as described earlier, an electrochemical cycling process was started in order to configure an atomic-scale switch, which allows reproducible bistable switching between an off-state and a well-defined quantized on-state. As soon as an upper threshold ($4.9G_0$) near the desired conductance value for the on-state ($5.0G_0$) was exceeded, the gate voltage was changed from a voltage in the deposition regime (+4 mV) to a voltage in the dissolution regime (-36 mV), the voltage being changed at a rate of 10 mV/s. As soon as the conductance dropped below a lower threshold of the source-drain conductance ($0.1G_0$), the gate voltage was changed back to a voltage within the deposition regime (+4 mV), again at a rate of 10 mV/s. This deposition process was continued until conductance exceeded the upper threshold of $4.9G_0$. At this point, a new cycle consisting of dissolution of the contact and subsequent deposition was started.

Figure 11.2b shows the conductance of the silver contact between the two gold working electrodes in units of the conductance quantum G_0 as a function of time

Fig. 11.2 Configuring a bistable atomic-scale switch by repeated electrochemical cycling. (**a**) Externally applied gate voltage as a function of time. (**b**) Corresponding change in contact conductance. Only after repeated cycling, regular switching is observed as a function of the applied gate voltage (see *arrow*). (**c**) Zoom-in into the data of (**b**): Regular switching of a bistable atomic-scale quantum point contact, induced by an applied gate voltage (cf. [1])

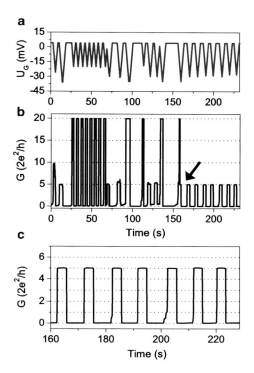

during this cycling process. Figure 11.2a gives the corresponding voltage applied to the gate electrode. As seen in Fig. 11.2b, during the first such dissolution/deposition cycles of each freshly-formed contact, conductance values strongly vary from cycle to cycle. In most cases, contact formation resulted in contact conductance values exceeding $20G_0$. When dissolving the contact, conductance immediately drops to zero.

During some of the cycles, however, deposition leads to the formation of a contact at a significantly lower conductance value. Yet, no reproducible response is observed as a function of the applied gate voltage. Not only the conductance observed by closing the gap, but also the time needed for forming and for dissolving a contact varies from cycle to cycle. While contact conductance values near $5G_0$ were observed several times, the behavior of the contact as a result of the applied gate voltage was still erratic in the beginning.

Only after 290 s from the beginning of the experiment (see black arrow in Fig. 11.2b), a sudden transition from an erratic to a regular behavior of the contact is observed. Beginning at this point, each of the following cycles of the gate voltage, as shown in Fig. 11.2a, results in a corresponding opening and closing of the gap, as shown in Fig. 11.2b, the conductance after closing the gap always being $5G_0$. A zoom-in into this sequence of regular switching events of Fig. 11.2b is given in Fig. 11.2c. Note that each cycle of the gate voltage results in the corresponding switching of the conductance between the "source" and "drain" electrodes: The device now reproducibly operates as an atomic-scale transistor.

This sudden transition from an irregular opening and closing of the contact to a bistable switching between zero and a well-defined quantized conductance value was also regularly observed for other electrochemically deposited silver point contacts, the on-state conductance commonly exhibiting values that were integer multiples of the conductance quantum G_0.

The fact that quantized conductance is observed at values of a few conductance quanta (e.g. $5G_0$ for the data shown earlier) means that the contact cross-section is still on the atomic scale [27,28]. The observation that reversible switching is found between two well-defined conductance values indicates that the contact switches between two well-defined configurations on the atomic scale. This could also explain the observed sudden transition between irregular and regular behavior of the source-drain conductance as a function of the gate voltage. As long as no atomic-scale bistability is formed, each deposition cycle leads to the electrochemical deposition of a new contact, different contacts having different conductance values. As soon as a bistable contact configuration has formed, the variation of the applied gate voltage could lead to a switching between the two contact configurations even before dissolution or redeposition of the contact happens. This switching does not necessarily involve electrochemical deposition and dissolution, but could also be induced by changes of local surface forces due to changes of the applied gate voltage: a variation of the gate voltage will lead to changes of the electrochemical double layer, which, in turn, will change surface forces and surface tension [29,30].

11.4 Preselectable Integer Quantum Conductance of Electrochemically Fabricated Silver Point Contacts

In practice, most conductance measurements of point contacts, even for simple metals, yield noninteger multiples of the conductance quantum G_0. Such deviations from the ideal behavior can stem from material-specific properties of the junction or from defects that result from the fabrication process. As shown earlier, our combined electrochemical deposition and annealing method for the fabrication of metallic quantum point contacts yields nearly ideal integer multiples of G_0 for the quantum conductance and explains their properties in comparison with conductance calculations for selected, near crystalline junction geometries with a preselected number of contact atoms.

Especially in experiments based on atomic-scale contact fabrication by mechanical deformation (e.g. break junctions or STM setups [12, 15, 16]), there is very limited control of the growth and properties of the atomic-scale contacts. In these experiments, long-term stable and defect-free contacts with conductance at integer multiples of the conductance quantum G_0 are difficult to realize in practice as the fabrication process is essentially connected with the formation of atomic-scale defects such as dislocations.

To produce well-ordered contacts, a technique of nearly defect-free growth by slow quasi-equilibrium deposition is required, which can be provided by electrochemical deposition methods [9, 17, 19, 20, 26]. Techniques of electrochemical annealing provide the possibility of healing atomic-scale defects in contacts even after fabrication (see the following section). Due to its high electrochemical exchange current density [31], silver is a promising candidate for efficiently applying electrochemical annealing techniques.

Now we discuss the experimental results of the electrochemical annealing method by electrochemical deposition/dissolution cycling of atomic-scale silver contacts in more detail and compare the experimentally observed conductance with the calculated conductance of modeled idealized junctions between two ideal single crystals with a predefined number of contacting atoms.

In the experiments, dissolution/deposition cycles between predefined conductance values were performed [1, 2]: Typically, after a number of cycles, a stable contact was formed, which exhibited an integer conductance value, and the cycling was stopped. Using this method, stable conductance levels at integer multiples of G_0 were configured. Examples of nG_0 ($n = 1, 2, 3, 4, 5$) are given in Fig. 11.3a. This

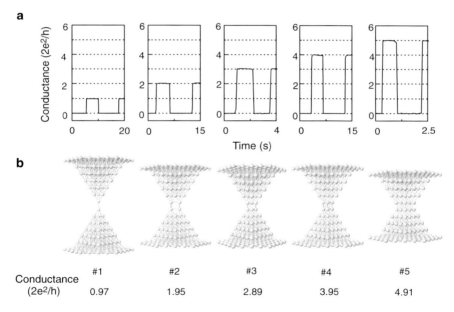

Fig. 11.3 Comparison of experimental conductance data of electrochemically annealed silver point contacts with calculations assuming idealized geometries. (**a**) Quantum conductance of five different annealed atomic-scale contacts at $1G_0$, $2G_0$, $3G_0$, $4G_0$, $5G_0$, respectively (with $1G_0 = 2e^2/h$), which were reversibly opened and closed. (**b**) Idealized geometries of silver point contacts with predefined numbers of contacting atoms. Conductance calculations performed within a Landauer approach result in near-integer multiples of G_0 for each of the five contact geometries (#1–#5). For the conformations shown above, the axis of symmetry of the junction corresponds to the crystallographic [111] direction. Reprinted with permission from [2]. © 2008 American Institute of Physics

transition from instable contacts with noninteger conductance to stable contacts with integer conductance values can be explained by an electrochemical annealing process, which heals defects in the direct contact region by electrochemical deposition and dissolution leading to an optimized contact configuration. After the electrochemical annealing process, most transitions appear to be instantaneous within the time resolution of the diagram of Fig. 11.3a (50 ms), whereas at higher time resolution (10 μs), fingerprints of the atomic-scale reorganization of the contact were observed in the form of both integer and noninteger instable transient levels.

In order to get insights into the possible structures of the measured point contacts, we calculated the coherent conductance of ideal crystalline silver nanojunctions (see Fig. 11.3b). Geometries were generated by assuming two fcc electrode clusters, which are connected at their tips by a small number of Ag–Ag-bridges in [111] direction with a bond length of 2.88 Å [32].

The zero-bias quantum conductance of a given junction geometry was computed with the Landauer formula [33, 34]. The electronic structure was described using an extended Hückel model [35, 36] including s- , p- , and d-orbitals for each silver atom (around 3,600 orbitals per junction). Consistently, material-specific surface Green's functions were computed using a decimation technique [34]. To reduce the influence of interference effects, we averaged the conductance, $G(E)$, over a small interval $[E_F - \Delta, E_F + \Delta]$ around the Fermi energy (with $\Delta = 50$ meV), which is comparable to the temperature smearing in measurements at room-temperature.

As shown in Fig. 11.3b, we find nearly integer conductance of the idealized geometries for contact geometries #1–#5 : $0.97G_0$, $1.95G_0$, $2.89G_0$, $3.95G_0$, and $4.91G_0$ respectively. The deviation from integer multiples of G_0 of about $0.1G_0$ is within the range of the accuracy of our numerical method. We observe a good correlation between the number of silver atoms at the point of minimal cross section and the number of conductance quanta, which aids in the construction of geometries with a particular value of the conductance. However, this is a material-specific property of silver not necessarily to be encountered in other materials.

Figure 11.4 shows the calculated transmission as a function of the electron energy within the energy interval $[E_F - 6\,\text{eV}, E_F + 6\,\text{eV}]$ for the five silver point contact geometries (#1–#5) given in Fig. 11.3b. The experimentally relevant values correspond to the conductance at the Fermi energy indicated by the vertical line in the figure. For the given silver junction geometries we obtained Fermi energies between -5.83 and -5.81 eV, which may be slightly below the correct value, caused by the known energy underestimation of the extended Hückel model [34]. The conductance curve oscillations are sensitive to the atomic positions. Therefore, an average of the conductance around the Fermi energy yields a more representative value of the conductance G, taking effectively into account the atomic vibrations during the measurement.

In order to study to which extent the conductance values change due to geometrical changes in the interatomic distance of the contacting atoms and the relative angle between the contacting crystals, we introduced finite changes in contact geometry: we calculated the electrode distance and twist-angle dependence of the zero bias conductance. Increasing the electrode distance to twice the Ag–Ag

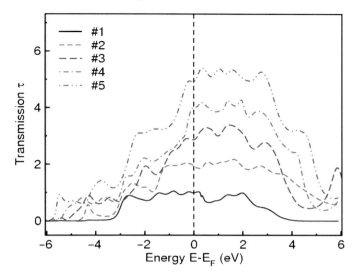

Fig. 11.4 Calculations of the transmission as a function of the electron energy for the five different silver contacts (#1–#5) of Fig. 11.3b. The experimentally relevant values correspond to the conductance at the Fermi energy indicated by the vertical line in the figure. Reprinted with permission from [2]. © 2008 American Institute of Physics

bond length leads to a decrease by 86.7% in the conductance, while twisting the electrodes by 60° against each other leads to a decrease of conductance by 22.0%.

11.5 Summary

The results demonstrate that for silver as a representative of a simple s-type metal, if defects and disorder in the contact area are avoided, the conductance in atomic-scale point contacts is typically an integer multiple of the conductance quantum G_0. The method of combined electrochemical deposition and electrochemical annealing of point contacts has proved to be a very efficient technique to generate such well-ordered contacts. On the other hand, if annealing is omitted, noninteger multiples of the conductance quantum are observed, which can be attributed to scattering due to defects and disorder within the contact area. These observations are confirmed by calculations on ideal model geometries of contacting silver nanocrystals, which yield integer multiples of the conductance quantum within the accuracy of the calculation in all the five cases investigated. As soon as disorder or local distortions of the atomic lattice within the contact area are introduced in the model geometry, drastic deviations from integer quantum conductance are obtained. This, in turn, indicates that such kind of disorder is effectively avoided in our experiments as a consequence of the electrochemical annealing approach. The results give an experimental proof of integer conductance quantization in annealed contact geometries of simple metals. The reproducible fabrication process also opens perspectives for the controlled configuration of atomic-scale quantum devices.

Acknowledgment The authors are thankful to Stefan Brendelberger for his experimental support. The SEM image was taken at the Laboratory for Electron Microscopy (LEM) of the Universität Karlsruhe. Fruitful discussions with Evgeni Starikov are gratefully acknowledged. The financial support was provided by the Deutsche Forschungsgemeinschaft within the Center for Functional Nanostructures (CFN), Projects B2.3/C5.1, and by Grant WE 1863/15–1.

References

1. F.-Q. Xie, Ch. Obermair, Th. Schimmel, in *Nanoscale Devices – Fundamentals and Applications*, ed. by R. Gross et al., (Springer, New York, 2006), p. 153
2. F.-Q. Xie, R. Maul, Ch. Obermair, E.B. Starikov, W. Wenzel, G. Schön, Th. Schimmel, Appl. Phys. Lett. **93**(4), 3103 (2008)
3. A. Nitzan, M.A. Ratner, Science **300**, 1384 (2003)
4. C. Joachim, J.K. Gimzewski, A. Aviram, Nature **408**, 541 (2000)
5. M.A. Reed, C. Zhou, C.J. Muller, T.P. Burgin, J. M. Tour, Science **278**, 252 (1997)
6. X.D. Cui, Science **294**, 571 (2001)
7. S.J. Tans, A.R.M. Verschueren, C. Dekker, Nature **393**, 49 (1998)
8. C.Z. Li, A. Bogozi, W. Huang, N.J. Tao, Nanotechnology **10**, 221 (1999)
9. F.-Q. Xie, L. Nittler, Ch. Obermair, Th. Schimmel, Phys. Rev. Lett. **93**, 128303 (2004)
10. K. Terabe, T. Hasegawa, T. Nakayama, M. Aono, Nature **433**, 47 (2005)
11. F. Xie, R. Maul, A. Augenstein, Ch. Obermair, E.B. Starikov, W. Wenzel, G. Schön, Th. Schimmel, Nano Lett. **8**(12), 4493 (2008)
12. N. Agraït, A. Levy Yeyati, J.M. van Ruitenbeek, Phys. Rep. **377**, 81 (2003)
13. N. Agraït, J.G. Rodrigo, S. Vieira, Phys. Rev. B **47**, 12345 (1993)
14. J. I. Pascual, Phys. Rev. Lett. **71**, 1852 (1993)
15. J.M. Krans, J.M. van Ruitenbeek, V.V. Fisun, I.K. Yanson, and L.J. de Jongh, Nature **375**, 767 (1995)
16. E. Scheer, Nature **394**, 154 (1998)
17. C.Z. Li, N.J. Tao, Appl. Phys. Lett. **72**, 894 (1998)
18. C.Z. Li, A. Bogozi, W. Huang, N.J. Tao, Nanotechnology **10**, 221 (1999)
19. A.F. Morpurgo, C.M. Marcus, D.B. Robinson, Appl. Phys. Lett. **74**, 2084 (1999)
20. C.Z. Li, H.X. He, N.J-Tao, Appl. Phys. Lett. **77**, 3995 (2000)
21. J. Li, T. Kanzaki, K. Murakoshi, Y. Nakato, Appl. Phys. Lett. **81**, 123 (2002)
22. F. Elhoussine, S. Mátéfi-Tempfli, A. Encinas, L. Piraux, Appl. Phys. Lett. **81**, 1681 (2002)
23. Ch. Obermair, R. Kniese, F.-Q. Xie, Th. Schimmel, in *Molecular Nanowires and Other Quantum Objects*, ed. by A.S. Alexandrov, J. Demsar, I.K. Yanson, (Kluwer Academic Publishers, The Netherlands, 2004), p. 233
24. D.M. Eigler, C.P. Lutz, W.E. Rudge, Nature **352**, 600 (1991)
25. H. Fuchs, Th. Schimmel, Adv. Mater. **3**, 112 (1991)
26. F.-Q. Xie, Ch. Obermair, Th. Schimmel, Solid State Commun. **132**, 437 (2004)
27. M. Brandbyge, K.W. Jacobsen, J.K. Norskov, Phys. Rev. B **55**, 2637 (1997)
28. J.C. Cuevas, A. Levy Yeyati, A. Martín-Rodero, Phys. Rev. Lett. **80**, 1066 (1998)
29. C.E. Bach, M. Giesen, H. Ibach, T.L. Einstein, Phys. Rev. Lett. **78**, 4225 (1997)
30. C. Friesen, N. Dimitrov, R.C. Cammarata, K. Sieradzki, Langmuir **17**, 807 (2001)
31. *Gmelin's Handbook of Inorganic Chemistry*, 8. edn. (Verlag Chemie, Weinheim, 1973), Silver, Part A4, p. 220
32. X. Guang-Can, J. Cluster Sci. **17**, 457 (2006)

33. J. Heurich, J.-C. Cuevas, W. Wenzel, G. Schön, Phys. Rev. Lett. **88**, 256803 (2002)
34. P. Damle, A.W. Ghosh, S. Datta, Chem. Phys. **281**, 171 (2002)
35. R. Hoffmann, J. Chem. Phys. **39**, 1397 (1963)
36. V. Rodrigues, J. Bettini, A.R. Rocha, L.G.C. Rego, D. Ugarte, Phys. Rev. B **65**, 153402 (2002)

Chapter 12
Electronically Tunable Nanostructures: Metals and Conducting Oxides

Subho Dasgupta, Robert Kruk, and Horst Hahn

Abstract Electric field-induced reversible tuning of physical properties, as opposed to property modification via irreversible variation in microstructure of materials, is discussed in this article. The foremost example of external field-controlled electronic transport of a material is the *"field effect transistors"* (FET)." However, the possibilities of tuning the macroscopic properties of materials with high charge carrier density have not been studied extensively. Large free carrier concentration in metals and high conducting oxides, however, can be of interest for specific applications. Despite the fact that the screening lengths of metals are extremely small, macroscopic property modulation can still be achieved via extremely small nanostructures (with very high surface-to-volume ratio). Moreover, electrochemical gating offers high surface charge density. While selected examples of tunable mechanical and magnetic properties of metals are cited, surface-charge-induced variation in electronic transport of metals (for both nanoporous and planar geometry) and high conducting transparent oxides are discussed in more detail.

12.1 Introduction

The tailoring of the mechanical, physical, and chemical properties of metals and alloys by modifying their microstructure is a well-known concept in material science. Large changes in material properties were observed through alloying or introducing point (dopants, vacancies, and interstitials), line (dislocations) and planar defects (grain and interphase boundaries). After the discovery of nanocrystalline and highly disordered materials, such as metallic glasses, large defect concentrations became a tool to tailor various physical properties, such as electrical, mechanical, catalytic, magnetic etc. Although sufficiently novel properties and consistent structure–property correlation were obtained, all of these property modulations via microstructural modifications are not reversible. In contrast, the modification of material properties by deviation from charge neutrality, e.g., by the application of a surface charge by an external applied field can be reversible and dynamic. High charge density per unit volume is the prerequisite to observe considerable change in the material properties. Consequently, the high surface-to-volume ratio of nanoparticles and nanoporous structures has been exploited. In conclusion, two modes of property modifications can be distinguished: (1) Tailoring of the microstructure resulting in irreversible changes of properties and (2) Tuning of the interfacial carrier density resulting in reversible changes of properties.

The mean field penetration, (δ), of an externally applied field inside a conducting solid essentially depends on the free carrier concentration. For a semiconductor with a carrier concentration of the order $10^{16}/cm^3$, field penetration or screening length can be as high as 100 nm (i.e. approx. 300 atomic layers) [1]. Consequently, orders of magnitude higher carrier concentration compared with the doping level or intrinsic carrier density is possible by externally applied field using a dielectric (e.g., SiO_2). This gives the working principle of normally-off junction field-effect transistor (JFET). In contrast, in case of metals, the carrier concentration is more than $10^{22}/cm^3$, which results in less than a monolayer screening length (0.25–0.3 nm for gold [2]). Therefore, the change in carrier concentration is negligible with a dielectric gating. However, a measurable change in physical properties can still be observed when a combination of a very high surface area of extremely small nanostructures and electrochemical charging is employed. A large surface charge density is possible with an electrolyte, as most metals possess a *double layer capacitance*, which is orders of magnitude higher than the maximum polarization possible with dielectrics. To ensure an electrochemical field-effect and negligible chemical doping (redox reaction at the interface), only the predominantly capacitive double layer region of a metal–electrolyte combination is utilized. Therefore, in all experiments, nonadsorbing supporting electrolytes are used. The *electrical double layer* at the metal–electrolyte interface is mostly analogous to a simple plate capacitor. Figure 12.1 shows the schematic of a capacitive double layer at a metal–electrolyte

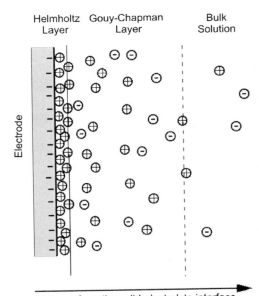

Fig. 12.1 Models of the electric double layer

interface and several models are presented to explain the counter ion concentration near the surface of contact.

Helmholtz model (1879): In this model, the double layer was mathematically treated as a simple capacitor with an immobilized layer of counter ions at the surface attached by electrostatic attraction, such that the surface charge is extremely neutralized. Therefore, the Helmholtz model predicts that the electric potential falls from its surface value to zero within the bulk solution over a thickness equal to the thickness of the counter ions attached to the surface.

Gouy–Chapman Model (1910–1913): In this model, the random thermal motion of the ions is considered, and therefore, if the counter ions are not chemically (or physically) adsorbed at the electrode surface, they cannot be immobilized. Therefore, it is suggested that the counter ions that neutralize the surface charge are spread out into the solution, forming what is known as diffuse double layer. According to this model, the surface potentially drops slowly to zero in the bulk solution over a large distance within the solution.

Stern Model (1924): This model is a combination of Helmholtz and Gouy–Chapman model. In this model, it is considered that there is indeed a layer of ions of one type (counter ions) near the electrode surface (Helmholtz layer), but the numbers are not enough to neutralize the charge. Hence the remainder of the charge is neutralized by a diffuse layer (Gouy–Chapman layer) extending out into the solution.

All these models essentially assume that

(a) Ions are effectively point charges
(b) The only significant interactions are Coulombic
(c) Electrical permittivity is constant throughout the double layer
(d) The solvent is uniform at the atomic scale

Besides the electrolytic charging, the other key to get significant change in macroscopic properties of metals or metallic conducting solids is the use of extremely small nanostructures. An exceptionally large surface-to-volume ratio with the number of surface atoms comparable with the bulk can be achieved with interconnected nanoparticle network or ultrathin films.

(a) *Ultrathin films*: The growth of continuous/percolating metal films is a challenge in itself. With the selection of the proper substrates (with a lattice matching for certain growth direction for the metal under consideration) and/or with a good choice of a buffer layer growth of percolating few nanometer thick metal films have been achieved. For instance, 2–4 nm thick FePt or FePd films on MgO substrate with Pt or Pd buffer layer have been reported in the literature [3] (Fig. 12.2).

(b) *Interconnected nanoparticles from compressed nanopowder synthesized by inert gas condensation*: Small particles of metals or alloys (with average particle size of a few nanometers) can be prepared by inert gas condensation, thermally evaporating the metal or the master alloy [4]. The agglomerates are loosely compressed to form a nonporous disc, consisting of an interconnected

Fig. 12.2 Schematic of the charged surface of a thin metal film inside electrolyte. Each individual sphere represents one atom

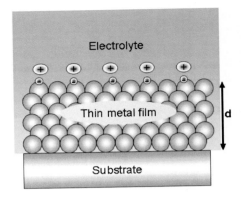

Fig. 12.3 Schematic of the charged surface of a nanoporous film/aggregate inside electrolyte. Each individual sphere represents one nanoparticle

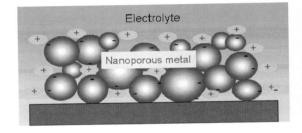

nanoparticle network with nanometer-sized pores, which are easily accessible by the electrolyte (Fig. 12.3).

(c) *Nanoporous network of noble metal from de-alloying*: De-alloying is a process of selective leaching, which refers to the selective removal of electrochemically more active metal from a master alloy. Typical examples of de-alloying could be dezincification (dissolving zinc from brass to get a porous copper structure), decarburization (graphitic corrosion from gray cast iron), or corroding silver out from a silver–gold alloy. The resultant nanoporous, spongy network is normally composed almost entirely of the electrochemically more noble metal. Although the physics behind the formation of nanopores is not entirely clear [5], it has been observed experimentally that the ligament (made of the nobler un-dissolved metal) and the pore size essentially depend on the applied de-alloying potential [6, 7].

As mentioned earlier in the manuscript, there are various physical properties that can be tailored reversibly through a deviation from charge neutrality by an external field.

(a) *Mechanical property*: Charge-induced reversible strain in the nanoporous metal compacts as a function of electrochemical surface charge has been studied by Weissmüller et al. [1, 8]. Piezoelectric response of the order of 0.1% or above was observed for nanoporous platinum when surface charge was applied with an aqueous solution of KOH [8]. Although the *electrocapilarity*, i.e. strain caused by an external voltage in an electrode due to electric double layer, has been well known since the nineteenth century, this was the first demonstration of a large strain in a pure metallic system. Later on, it has also been shown that a large macroscopic bending of a bimetallic metal-nanoporous metal cantilever of the order of several millimeters is possible with the same basic principle [9].

(b) *Magnetic property*: Magnetization of 3-D transition metals arises from partly filled d-band. It was proposed [1, 10] that an external field can reversibly tune the d-band state of a ferro- or paramagnetic metal or alloy resulting in a change in magnetization. Indeed a 1% variation in magnetic susceptibility is observed for paramagnetic nanoporous Pd upon electrochemical charging [11]. Recently, a charge-induced variation of coercivity and magnetic anisotropy was reported for ultrathin Pt–Fe and Pd–Fe films [3]. This can be a more convincing proof of the change in d-band filling in a ferromagnet with respect to an applied electric field.

(c) *Electrical Property*: The most renowned and technologically important example of controlled enhancement or depletion of the local charge carrier density by an external applied field is the '*junction field-effect transistor* (JFET)'. The device principle is to modulate the current flowing through a semiconductor by an external field applied by a gate and a dielectric. In case of an n-channel JFET, an enhancement or a depletion of the channel conductance (current through the semiconductor) is possible by a positive or negative gate bias respectively. It might be possible in a similar way to vary the conductivity of a pure metal or a near metallic system with higher local electric field at the surface. To tune the conductance of a metallic system, large surface-to-volume ratio of the nanostructures and an electrochemical gating was used. A reversible change in resistance of $\pm 4\%$ of a nanoporous compact of Pt nanoparticles with an aqueous electrolytic gating was reported recently [12].

The major advantage of using electrochemical charging to gate highly conducting, unconventional systems lies in the fact that

(a) an electrolyte can apply very high surface charge with a local field in excess of $10\,\text{MV/cm}^2$, up to two orders of magnitude higher than any existing gate dielectrics.

(b) For a nanoporous structure with interconnected nanoparticles, the use of electrochemical gating is even more beneficial as an electrolyte can interpenetrate the nanopores of the porous compact to surround all the nanoparticles from all sides. Therefore, each particle experiences the external field and surface charge from all sides as opposed to a dielectric gating where the field is applied only from one side and the particles far from the gate oxide interface hardly sense the applied field.

12.2 Tunable Change in Electronic Transport of a Metal

12.2.1 Nanoporous Gold Electrode from De-alloying

De-alloying of the polycrystalline master alloy of composition $Au_{0.25}Ag_{0.75}$ was performed in 1 M perchloric acid with a de-alloying potential of 0.75 V with respect to Ag/AgCl reference electrode. The as-prepared de-alloyed nanoporous gold (NPG) sample shows a very small ligament size of less than 5 nm, which undergoes considerable coarsening when immersed into a nonaqueous electrolyte ($LiClO_4$ in ethyl acetate) to measure the charge-induced modulation of conductivity.

The initial coarsening process was monitored by the decrease of resistance of the NPG sample. During the coarsening process, a decrease in resistance more than five times was observed. After the coarsening and the corresponding resistance change stopped to alter with time, the potential of the NPG was varied between −0.4 and 0.5 V with respect to Ag/AgCl reference electrode with a step size of 0.1 V. At the same time, a constant current of 1 mA was passed through the sample and the potential drop was measured. A reversible change in resistance of 6% with respect to the applied surface charge was obtained [13]. Due to the complicated microstructure with very small ligament sizes, the accurate measurement of the surface area and hence the surface charge density was not possible. However, the knowledge of the surface charge density is the prerequisite to have a quantitative model toward the understanding of the observed phenomena. To have a better control over the geometrical conditions, the planar geometry of thin films is ideal. Therefore, to understand the effect of different contributions like change in carrier concentration and/or a change in surface scattering, on the observed variation in resistance similar measurements were carried out with ultrathin gold film electrodes.

12.2.2 Variation in Resistance in Thin Gold Film Electrode

The change in resistance of the nanostructured metals with respect to an applied field is believed to be due to a change in carrier concentration and hence a linear variation of resistance with the surface charge is expected. Furthermore, in the earlier studies [12, 14, 15], the surface charge-induced change in resistance has been explained in light of the Drude model assuming that the resistance change is only due to the change in carrier concentration upon charging.

The free-electron gas model (Drude model) [16]:

$$\sigma = \frac{ne^2\tau}{m} \quad (12.1)$$

Hence,

$$\frac{\Delta R}{R} = -\frac{\Delta n}{n} \quad (12.2)$$

Fig. 12.4 Schematic representation of the experimental setup used for the charge-induced resistance modulation measurements of thin gold films

where σ is the conductivity; n, number of free-carriers (electrons); τ, relaxation time; R, resistance; and e and m are the charge and mass of the electron respectively. However, we have observed that in case of nanoparticulate systems the accurate estimation of surface area is highly challenging, which often in turn results in an overestimation of surface charge density. Therefore, careful measurements were carried out with thin gold films having a thickness of several nanometer (7, 9.3, 11.6 nm calculated from X-ray reflectometry). Figure 12.4 shows the experimental setup. The gold film was sputtered through a shadow mask to achieve an appropriate Van der Pauw geometry. A liquid electrolyte of 0.1 M $NaClO_4$ in propylene carbonate is used. An adsorption-free potential window is selected through ultraslow cyclovoltammogram (with a scan rate of 5×10^{-4} V/s). The potential of the gold film was varied between $+0.25$ and -0.25 V with respect to a Pt pseudo-reference electrode. At the same time, a constant current of 1 mA was passed through the gold film and the potential drop was recorded. The resistance measurement was done by the Van der Pauw method and noticeably a higher effect size was obtained compared with the expectations from change in free-carrier density (change in free-carrier density can be calculated by integrating the charging current, $\Delta C = e \, \Delta n$).

Therefore, a new model was proposed [17] to explain the observed phenomenon with the following conditions:

(a) The electron density profile of a charged metal surface shifts toward or away from the metal–electrolyte interface with respect to a positive/negative surface charge respectively [18]. This shift in the center of mass of the electron density profile can be viewed as a change of the effective thickness of the film.

(b) A change in scattering cross-section is expected due to this change in effective thickness according to the Fuchs–Sondheimer model [19, 20].

A self-consistent calculation of the shift in electron density profile for a strongly charged metal surface is given by Gies and Gerhardts [18]. Suitable values of the specularity parameter ($p = 0.5$) and the grain boundary reflection coefficient ($R = 0.85$) for gold thin films were taken from the literature [21, 22]. The selected values seem to be appropriate, when cross-checked with the experimental results. The experimentally obtained change in resistivity values for different film thickness matches well with the change expected from grain and surface scattering calculated using Mathiessen's rule [21].

Therefore, combining both of the contributions mentioned earlier, we get,

$$\frac{\Delta R_S}{R_S} = \frac{\partial \rho}{\partial t} \frac{\Delta t}{\rho} - \frac{\Delta t}{t} \quad (12.3)$$

where R_S is the sheet resistance; ρ, the resistivity; t, the film thickness; and Δt is the effective change in film thickness (shift of the center of mass of the electron density profile). In Fig. 12.5, the experimentally obtained relative change in resistance ($\Delta R_S/R_S$) is plotted with the theoretically calculated ones.

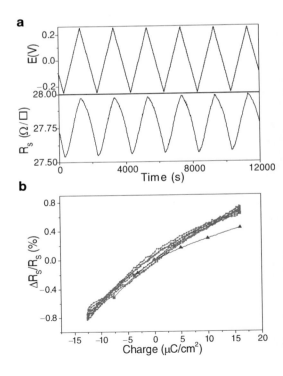

Fig. 12.5 (a) Change of the sheet resistance with respect to applied potential for 7 nm Au film, (b) Change of the resistance vs. charge (*filled circles*), calculated with Δt (σ) from Theophilou and Modinos (*filled triangle*), calculated with Δt (σ) from Gies and Gerhardts work (*filled squares*)

The calculated curves (wine and red) are computed using (12.3) with the values of the shift of density profile Δt, taken from the work of Theophilou and Modinos [23] and Gies and Gerhardts [18] respectively.

It can be noted here, that the shifts of electron density profile toward or away from the metal surface are not of the same magnitude when the same surface charge of the opposite sign is applied. The inward movement of the center of mass for a positively charged surface experiences a repulsive force outwards due to the large electron mass inside the metal, whereas the absence of this repulsive force for a negatively charged surface results in an easy spreading of the electron density profile. Therefore, a different response of the sheet resistance can be expected for different signs of the surface charge. Consequently, it can be concluded that a nominal nonlinear response of the resistance of a metal with respect to an electrochemically applied surface charge does not necessarily indicate an onset of a redox reaction.

12.3 Reversible Change in Electronic Transport in a High Conducting Transparent Oxide Nanoparticulate Thin Film

Indium tin oxide (ITO) is a degenerated semiconductor with a carrier concentration as high as $10^{21}/cm^3$ and conductivity as low as $10^{-5}/\Omega\,cm$. In this study, a dispersion of 5 at.% Sn-doped indium oxide nanoparticles with an average particle size of 15 nm is used. ITO is a stable oxide and therefore shows considerable large adsorption-free electrochemical window when immersed into a nonadsorbing electrolyte.

The sample geometry is shown in Fig. 12.6. The contact leads are made of 310 nm thick ITO films (dark in Fig. 12.6) on a glass (bright in Fig. 12.6) substrate. The device fabrication starts with a 310 nm thick sputtered ITO film on high quality float glass. The leads are prepared using e-beam lithography on spin-coated PMMA and ion etching. Subsequently, a second lithography step to create a $8 \times 2\,\mu m^2$ window for spin coating of the ITO nanoparticles dispersion (Evonik Degussa GmbH) follows. After the spin coating of the nanoparticle dispersion,

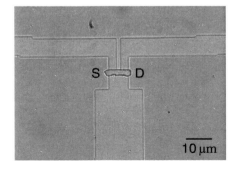

Fig. 12.6 Optical image of the device showing the source, drain made of sputtered ITO film and the channel made of ITO nanoparticles

a low-temperature (185°C) annealing for 1 h, prior to the PMMA layer lift-off with acetone, is performed. The whole structure is then annealed in air for 2 h at 500°C to eliminate the organic surfactants from the ITO surfaces. No significant particle growth is observed to occur during the annealing in air, which has also been observed by other authors for ITO nanoparticles [24, 25].

An electrolyte consisting of 0.1 M NaClO$_4$ in propylene carbonate is used. High surface area (>1000 m^2/g) activated carbon cloth (kynol) is used as a counter and a pure (99.99%, Chempur) platinum wire is used as a pseudo-reference electrode.

All the measurements are carried out at room temperature. The cyclovoltametry was performed with a glass substrate entirely spin coated with ITO nanoparticles to determine the potential window with negligible adsorption. Figure 12.7a shows the cyclovoltammogram with a scan rate of 5×10^{-4} V/s. The accumulated charge is determined by integrating the charging current. The identical potential window with the same scan rate is used during the measurement of the resistance change of the ITO nanoparticle device. A constant current of 0.1 μA is applied and the potential drop across two terminals is measured. More than two orders of magnitude change in resistance (on/off $= 3.25 \times 10^2$) is observed when the potential of the working electrode is varied between -0.25 and 0.85 V ($V_g = -0.85$ to 0.25 V) (Fig. 12.7b).

At the negative potential of the working electrode (positive gate potential), positive ions (i.e. Na$^+$ ions) come close to the surface of the ITO nanoparticles. As a result, the electron charge redistribution in the ITO particles occurs and electrons are attracted toward the surface to build the charge double layer. Therefore, a negative electrode potential (positive gate potential) increases the carrier (electron) density of the ITO nanoparticles resulting in a decrease of the resistance of the channel. In case of a positive electrode potential (negative gate), negative ions (i.e. ClO$_4^-$) form the charge double layer, which repels the electrons away from the metal–electrolyte interface. Accordingly, the electron density of the channel decreases with a corresponding increase of the resistance.

Therefore, the positive or negative ions of the charge double layer at the solid–electrolyte interface work as a gate and the channel gets narrower and broader with the negative and positive gate voltage, respectively, making the device analogous to a normally-on junction field-effect transistor (JFET). Figure 12.7c shows the response of the device with a potential pulse of 0.25 and -0.85 V applied after 180 s. The resistance of the channel is measured with 25 ms time intervals. Figure 12.8 shows the drain current–drain voltage characteristic of the device. The gate voltage is varied between 0.2 and -0.8 V, while the drain voltage is varied between 5 mV and a maximum 980 mV.

Figure 12.9 shows the drain and source current for the 20–10 mV drain voltage, which shows that the leakage current is more than two orders of magnitude lower than the drain/source current for all the gate and drain voltage combinations. Another important device performance that leads to a low power operation is the subthreshold swing (S). A small subthreshold swing is favored for switching transistors, while the theoretical limit is 60 mV/decade. The *subthreshold swing* observed for the present device is 415 mV/decade. This type of system could represent a valuable contribution to the printable/flexible electronics community, particu-

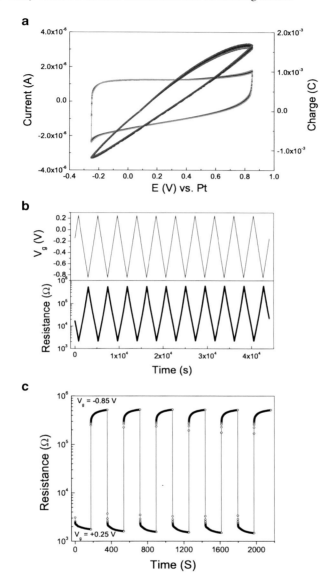

Fig. 12.7 (a) Cyclovoltammogram of dispersed ITO nanoparticles. (b) Change in resistance of the ITO channel when the gate voltage is varied with a constant rate of 5×10^{-4} V/s. (c) Change in the channel resistance in response of potential pulses between the gate and channel applied after 180 s

larly for the growing number of researchers who are switching their efforts in this area away from traditional, heavily explored organics to new classes of materials, including nanostructured inorganics. An all-solid-state device based on the same basic principle, prepared using a polymeric solid electrolyte, can be

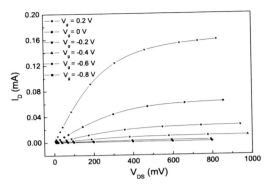

Fig. 12.8 The drain current–drain voltage characteristic of the device when the gate voltage varied between +200 to −800 mV

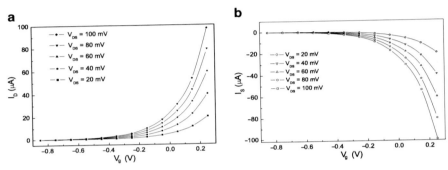

Fig. 12.9 (**a**) The drain current and (**b**) source current measured at different gate (V_g = 250 to −850 mV) and drain voltage (V_d = 20 to 100 mV)

appropriate for the engineering developments toward easy-to-fabricate and printable macroelectronics [26].

12.4 Summary

Reversible tuning of physical properties of nanostructures with respect to externally applied electric field is discussed. Very high surface charge density is achieved through *electrochemical double layer* charging. Despite the high free-carrier density and extremely low mean field penetration, a change in resistance of a few percentages (6%) was measured for nanoporous gold film. A new and complete explanation of the phenomenon is proposed based on the experimental result on thin gold films. It is shown that several orders of magnitude higher effect size is possible when the nanostructured pure metal is replaced by a nanoporous conducting oxide. The larger screening length of a conducting oxide, such as ITO, as a result of two orders of magnitude lower electron density, results in an effect size in the range required for potential applications.

Acknowledgment The authors are thankful to Simone Dehm for her help with e-beam lithography. The fruitful discussions with Dr. F. Evers and Dr. J. Weissmüller are gratefully acknowledged. The financial support was provided by the Deutsche Forschungsgemeinschaft (DFG) and the State of Baden-Württemberg through the DFG-Center for Functional Nanostructures (CFN) within subproject D4.4. The financial support by the State of Hessen is also appreciated.

References

1. H. Gleiter, J. Weissmüller, O. Wollersheim, R. Würschum, Acta Mater. **49**, 737 (2001)
2. K. Kempa, Surf. Sci. **157**, L323 (1985)
3. M. Weisheit, S. Fähler, A. Marty, Y. Souche, C. Poinsignon, D. Givord, Science **315**, 349 (2007)
4. C. Lemier, S. Ghosh, R.N. Viswanath, G.-T. Fei, J. Weissmüller, Mater. Res. Soc. Symp. Proc. **876E**, R2.6.1 (2005)
5. A.J. Forty, Nature **282**, 597 (1979)
6. S. Parida, D. Kramer, C.A. Volkert, H. Rösner, J. Erlebacher, J. Weissmüller, Phys. Rev. Lett. **97**, 035504 (2006)
7. J. Erlebacher, J. Electrochem. Soc. **151**, C614 (2004)
8. J. Weissmüller, R.N. Viswanath, D. Kramer, P. Zimmer, R. Würschum, H. Gleiter, Science **300**, 312 (2003)
9. D. Kramer, R.N. Viswanath, J. Weissmüller, Nano Lett. **4**, 793 (2004)
10. H. Gleiter, Scripta Mater. **44**, 1161 (2000)
11. H. Drings, R.N. Viswanath, D. Kramer, C. Lemier, J. Weissmüller, R. Würschum, Appl. Phys. Lett. **88**, 253103 (2006)
12. M. Sagmeister, U. Brossmann, S. Landgraf, R. Würschum, Phys. Rev. Lett. **96**, 156601 (2006)
13. A.K. Mishra, C. Bansal, H. Hahn, J. Appl. Phys. **103**, 094308 (2008)
14. R.I. Tucceri, D. Posadas, J. Electroanal. Chem. **191**, 387 (1985)
15. R. I. Tucceri, D. Posadas, J. Electroanal. Chem. **283**, 159 (1990)
16. N.W. Ashcroft, N.D. Mermin, *Solid State Physics* (Holt-Saunders International Edition, Japan, 1981), p. 7
17. S. Dasgupta, R. Kruk, D. Ebke, A. Hütten, C. Bansal, H. Hahn, J. Appl. Phys. **104**, 103707 (2008)
18. P. Gies, R.R. Gerhardts, Phys. Rev. B **33**, 982 (1986)
19. K. Fuchs, Proc. Camb. Philos. Soc. **34**, 100 (1938)
20. E.H. Sondheimer, Adv. Phys. **1**, 1 (1952)
21. C. Durkan, M.E. Welland, Phys. Rev. B **61**, 215 (2000)
22. M.A. Schneider, M. Wenderoth, A.J. Heinrich, M.A. Rosentreter, R.G. Ulbrich, Appl. Phys. Lett. **69**, 1327 (1996)
23. A. Theophilou, A. Modinos, Phys. Rev. B **6**, 801 (1972)
24. J. Ederth, P. Johnsson, G.A. Niklasson, A. Hoel, A. Hultåker, P. Heszler, C.G. Granqvist, A.R. van Doorn, M.J. Jongerius, D. Burgard, Phys. Rev. **B 68**, 155410 (2003)
25. J. Ederth, G.A. Niklasson, A. Hultåker, P. Heszler, C.G. Granqvist, A.R. van Doorn, M.J. Jongerius, D. Burgard, J. Appl. Phys. **93**, 984 (2003)
26. S. Dasgupta, S. Gottschalk, R. Kruk, H. Hahn, Nanotechnology **19**, 435203 (2008)

Chapter 13
Impedance Spectroscopy as a Powerful Tool for Better Understanding and Controlling the Pore Growth Mechanism in Semiconductors

A. Cojocaru, E. Foca, J. Carstensen, M. Leisner, I.M. Tiginyanu, and H. Föll

Abstract This work shows new results toward a better understanding of macropore growth in semiconductors by using in situ FFT impedance spectroscopy. A new interpretation of the voltage impedance is proposed. In particular, the pore quality could be quantified for the first time in situ, especially by extracting the valence of the electrochemical process. The study paves the way toward an automatized etching system where the pore etching parameters are adjusted in situ during the pore etching process.

13.1 Introduction

The electrochemical pore formation in Si is a topic [1] with many potential applications, and much progress was made toward the development of production technologies [2] and many product prototypes were advanced (see [3] and references therein). Despite all of this work, no product based on porous Si can be found on the market at present. Among the main reasons for this is the still-not-fully-understood mechanism of pore formation, or more generally, the many open questions in the field of electrochemistry of semiconductors. As an example, many envisioned applications demand precise control of the pore quality (e.g. diameter variations, pore wall roughness) and the present understanding of pore formation mechanisms, although rather advanced in some respects, does not ensure full control of the etching process as would be needed.

In this work, we show how impedance spectroscopy can be used for the purpose of controlling the macropore growth in n-type Si. In particular, it is shown how one can extract the dissolution valence at the pore tips from the measured impedance. This number is used as a quantification of the pore quality. Determining this number in situ can pave the way toward the implementation of an automatized etching system.

13.2 Experimental

n-Type Si wafers with low doping levels corresponding to a resistivity of 5 Ω cm are used for etching macropores. The substrate orientation is (100) with an n^+ layer on the backside of the wafer for a good ohmic contact to the sample. The etching is done using backside illumination (BSI) [4]. The samples were prestructured by standard photolithography before etching; the nucleation pattern was a hexagonal lattice with a lattice constant of $a = 4.2\,\mu$m. The electrolyte consisted of 5 wt.% HF in an aqueous electrolyte. The temperature of the electrolyte was fixed at 20°C. The FFT impedance spectrometer embedded within the etching system provided by the ET&TE GmbH, Germany was used to extract information concerning pore growth during the etching process.

13.3 Results and Discussion

For voltage impedance, a small perturbation signal is applied to the anodization voltage and the response in the etching current is measured. Figure 13.1a shows the $I-V$ curve of n-type Si in contact with HF under BSI. One can easily see that a change in the voltage causes a variation in the current. Since during the macropore etching the etching voltage must be in the saturation regime of the $I-V$ curve [2], the linearity condition is fully fulfilled. However, another problem can be seen, i.e., being in the saturation regime, any perturbation in the voltage generates a minute variation of the current. In order to separate the measured signal from noise, strong requirements are imposed to the measuring hardware as well as to the data processing software.

Figure 13.1b shows a typical Nyquist plot of the measured impedance. The squares indicate the measured data. The data are fitted using the following model:

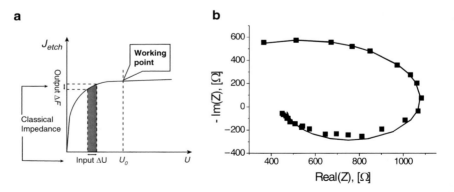

Fig. 13.1 (a) The IV curve of the n-type Si/HF interface with BSI. (b) The Nyquist plot of the measured voltage impedance together with the fitting curve

$$Z(\omega) = R_s + \cfrac{1}{\left(\cfrac{i\omega\tau}{(R_p+\Delta R_p)(1+i\omega\tau)} + \cfrac{1}{R_p I(1+i\omega\tau)}\right) + i\omega C_p} \quad (13.1)$$

where R_s is the serial resistance accounting for the voltage losses at the contacts and electrolyte, R_p and ΔR_p are the chemical transfer resistances, which in our model are frequency dependent, τ is a time constant, and C_p is the parallel capacitance describing the capacitive nature of the reaction interface. A full derivation of this model can be found elsewhere [5]. The solid fitting curve in Fig. 13.1b shows a very good match between the fit model and the measured impedance.

Generally, interpreting the impedance data for macropore formation under BSI for all kinds of etching conditions is nearly impossible, since a lot of electrochemical dissolution processes can occur in parallel, e.g. the (desired) photoinduced dissolution of Si at the pore tips, the (undesired) dissolution of Si at the pore walls by electrical breakthrough current, or the formation of side pores. Therefore, the voltage impedance can look much more complicated than for the formation of "nice" macropores as described in this paper.

While the capacitive behavior of the measured impedance during macropore formation could be well described by other authors [6], most challenging is the measured "inductive" loop. As mentioned earlier, R_p and ΔR_p, which represent the intersection points of the inductive loop with the abscissa axis, are related to the chemical processes that take place at the reaction interface. Under the optimal pore growth conditions this is solely represented by the pore tip, the main reactions at the pore tip, involving charge transfer, represent the Si dissolution. We claim that the R_p and ΔR_p represent the dissolution reactions, which are slow, i.e., R_p for $\omega \to 0$, and fast, i.e., $R_p + \Delta R_p$ for $\omega > 0$, respectively. Hence, the parallel resistance is determined by two processes with different reaction rates: fast and slow. We assign this to the two well-known and accepted main reactions for the dissolution of Si: divalent (fast, the direct process of Si dissolution) and tetravalent (slow, purely chemical dissolution of the oxide in order to dissolve Si).

Without going to much into details, here (more can be found in [5]), the measured R_p and ΔR_p can be used to calculate a quantity n, which describes very well the quality of the obtained macropores. It is defined as:

$$n = \cfrac{4}{2 - \cfrac{\Delta R_p}{R_p + \Delta R_p}} \quad (13.2)$$

Analyzing the time behavior of n, we concluded that it represents the dissolution valence at the pore tip. By using the voltage impedance, this magnitude can be measured in situ. It is important to mention that the dissolution valence was earlier measured via gravimetry and a value of 2.7 was found [7]. As it will be further shown, the measured values in this work fit very well with these numbers.

Figures 13.2a, b show two types of pores etched under nonoptimized conditions: in (a) macropores are etched with a small diameter such that the pore walls are fully conductive, i.e., the photogenerated holes can penetrate the pore interspacing and be

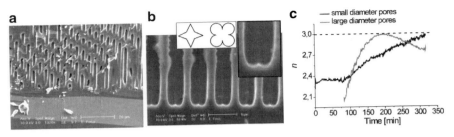

Fig. 13.2 SEM cross section of macropores etched under unoptimized conditions leading to (**a**) too large interspacing between the pores or (**b**) an extremely large diameter and star- or petal-like shape of the macropores. (**c**) The measured dissolution valence for both cases

consumed in electrochemical reactions at the pore walls; in (b) pores are shown that were etched with a very large diameter such that the pore walls are insulating; later, however, the anodization voltage is increased such that the pores become star or petal-shaped with features that, due to their sharp geometries, advantage the current flow.

Figure 13.2c shows the calculated dissolution valence from the measured impedance data. Striking is the value of 3, which for the case of small diameter pores is hardly reached, while the large-diameter pores reach this value relatively fast. However, after almost 200 min, the curve decreased again.

The maximum value of the valence can be 3 if the electrochemical reactions would take place exclusively at the pore tip, or it reduces to 2 if n-type Si is etched in the dark [2]. Since the measured valence is an average describing the dissolution at the pore tip as and the (unwanted, but unavoidable) dissolution at the pore walls, it becomes obvious why the values in Fig. 13.2c are situated between 2 and 3. Obviously, the small diameter pores, where the holes penetrate easily between the pores and thus increase the leakage current, exhibit an average valence well below 3, which is caused by the strong component of the divalent dissolution. In contrast, when large-diameter pores are etched, due to completely insulating pore walls, the measured valence reached 3 very fast, however decreasing toward 2 once the voltage is too high and the macropores become star or petal-shaped, hence increasing the leakage current via the pore walls, i.e., the divalent dissolution.

For etching homogeneous macropores the leakage current, i.e., the divalent dissolution of the pore walls might be very small. Measuring the dissolution valence in situ might help to assess the moment when leakage current gets large and eventually to try to avoid it by readjusting the etching parameters. Fig. 13.3 shows such an example. The SEM picture in Fig. 13.3a shows the cross section through a porous structure that at the first glance seems to be etched perfectly. However, a closer inspection reveals that there are regions on the sample of badly grown macropores, in particular, where the diameter is not constant and also the lattice is destroyed. The measured valence, according to Fig. 13.3b, shows a fast increase to 3, indicating a good start of the etching. After 300 min, the dissolution valence starts to decrease toward 2, indicating that the leakage current through the walls becomes dominant

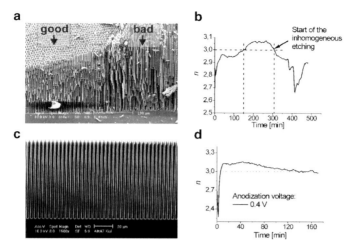

Fig. 13.3 (a) SEM image in cross section which shows good and bad macropores on the same sample. (b) The extracted dissolution valence for the structure in (a). (c) SEM picture of the sample in cross section where the pore growth did function well and (d) the extracted dissolution valence from the impedance data

and the porous structure is compromised. Figure 13.3c illustrates an SEM picture in cross-section through a sample that was etched perfectly. The very homogenous macropores and good pore quality can easily be observed. The valence reaches a value of 3 very fast where it remains for longer time.

Without impedance analyses during the pore etching, the only way to check whether the pores grow perfectly is the analysis at the illumination intensity as a function of time. For the sample shown in Fig. 13.3a, the illumination intensity gave signs for problems in the macropores etching only after 400 min. By means of the impedance, however, already at the 250 min, it could be observed that the etching process is not under optimized condition. Having the valence as a quality monitor tool allows at an early stage the readjustment of the etching parameters, which could easily save the etched structure. After 400 min, no parameter adapting allows to reestablish the normal etching process and the porous structure is considerably destroyed. Hence, the voltage impedance becomes a very powerful tool that can be effectively used in order to control in situ the etching process.

13.4 Conclusions

The data extracted from the voltage impedance allow to estimate the dissolution valence at the pore tip at any instant of time. This is for the first time that such precise information about the development of the pore tip state could be obtained. By comparing the valence with the quality of the pores, it could be concluded that the optimum pore growth takes place at a valence of $n \sim 3$. This justifies the assumption that the pores advance while producing, and ulterior dissolving, oxide at the pore tips. As a next step, a special model is to be developed for the purpose of explaining the growth mechanism of the macropores under optimized conditions.

Acknowledgments Parts of this work have been supported by the Alexander von Humboldt Foundation.

References

1. A. Uhlir, Bell System Tech. J. **35**, 333 (1956)
2. V. Lehmann, *Electrochemistry of Silicon* (Wiley-VCH, Weinheim, 2002)
3. L.T. Canham, A. Nassiopoulou, V. Parkhutik (eds.), Phys. Stat. Sol. (a) **202**(8) (2005)
4. V. Lehmann, H. Föll, J. Electrochem. Soc. **137**, 653 (1990)
5. E. Foca, PhD-Thesis, Christian-Albrechts-University of Kiel, 2007
6. J.-N. Chazalviel, Electrochim. Acta **35**, 1545 (1990)
7. S. Rönnebeck, J. Carstensen, S. Ottow, H. Föll, Electrochem. Solid State Lett. **2**, 126 (1999)

Chapter 14
Studying Functional Electrode Structures with Combined Scanning Probe Techniques

P. Dupeyrat, M. Müller, R. Gröger, Th. Koch, C. Eßmann, M. Barczewski, and Th. Schimmel

Abstract In this article, the study of the active electrode surfaces of a special type of fuel cells, solid oxide fuel cells (SOFCs), is described. The high conversion efficiency of this type of fuel cells is connected with a high working temperature, which makes the material selection difficult and causes thermal degradation processes, limiting the life time of the cells. We investigated the topography, conductivity phenomena, and chemical composition of such electrode surfaces in the nanoscale regime with several atomic force microscopy (AFM)-based techniques. Of particular interest was the grain size distribution of the surfaces in order to optimize their production process. The results of the AFM experiments were compared with those obtained by XRD measurements. The AFM achieved grain height data inaccessible by SEM. Nanoscale material contrast was obtained by applying the technique of Chemical Contrast Imaging (CCI) developed in our group. These measurements indicated possible phase separations in the electrode surfaces. The local electronic conductivity and differences between crystallites and grain boundaries were studied by Conductive AFM (C-AFM) and Electrostatic Force Microscopy (EFM). The theory, implementation and testing of this method and the results are discussed.

14.1 Introduction

Fuel cells are regarded as a high-potential alternative or replacement for conventional electrical generators. A special type of fuel cells with high conversion efficiency are solid oxide fuels cells (SOFCs). However, their high conversion efficiency is connected with a high working temperature in the range of about 700–1,000°C, which is needed to achieve a sufficiently high ionic conductivity and low electrode resistance. The high working temperature leads to difficulties regarding the material selection, to thermal degradation processes, and a complicated thermal management. All of this limits the long-term life time of SOFCs.

One of the key issues in further developing the SOFCs is the control and optimization of the electrode properties. Especially electrode materials with ionic and mixed ionic–electronic conductors (MIEC) are of interest. The increased ionic and electronic conductivity of nanostructured MIEC electrodes could allow lower operating temperatures leading to improved long-term stability, faster start-up times, and less complicated thermal management.

Apart from the choice and development of suitable electrode materials, the introduction of a nanostructured, polycrystalline interface layer between the cathode and the electrolyte could potentially improve the performance of SOFCs [1, 2]. Such interface layers can be produced, e.g., by metal organic deposition (MOD), which is a simple, inexpensive, and easily implementable coating method. The exploration of the structure and conductivity of the MOD layers are crucial issues for the improvement of electrode materials. That is due to the expectation that nanostructured cathodes will have a substantially increased ionic conductivity by accelerated oxygen diffusion at lower temperatures, because of their high density of grain boundaries [3].

This work was therefore aimed at the nanoscale investigation of the topography and nanoscale mapping of chemically different surface areas and conductivity phenomena with atomic force microscopy (AFM)-based techniques. Several AFM techniques – only one available in our group – have been used for the detection of the following material properties:

1. The topography – particularly the grain sizes and heights – of the MOD layers by AFM in order to be able to optimize their roughness and grain sizes.
2. Nanoscale material contrast on MOD surfaces by applying the technique of chemical contrast imaging (CCI) developed in our group [4–7].
3. The detection of the local electronic conductivity and properties on MOD surfaces and differences between crystallites and grain boundaries by electrostatic force microscopy (EFM) and conducting-AFM (C-AFM).

With topography measurements on 8YSZ-MOD layers (YSZ, yttria stabilized zirconia) prepared with different annealing temperatures, a strong correlation between the grain size distribution and annealing temperature could be observed. The surfaces of the MOD layers are found to be homogeneous. The results correspond well to those achieved by XRD. Furthermore, the AFM measurements resulted in grain height data inaccessible by SEM measurements (Tables 14.1 and 14.2).

Indications of a phase separation in the LSC50-MOD layers have been observed in another work. Measurements with CCI can show material contrast if the different phases lead to a different tribological interaction between the AFM tip and surface. Our measurements with CCI did show possible indications of phase separations in the LSC50-MOD layers in the range of the grain size.

Table 14.1 Average values for grain size and grain height of 8YSZ-MOD layers (MOD 39–MOD 43) for different temperature treatments

Temperature (°C)	Grain size (nm)	$\sigma_{\text{Grain Size (nm)}}$	Height (nm)	σ_{Height} (nm)
500	<8	–	<2	–
1,000	48	15	15	9
1,250	530	120	45	15
1,350	710	130	65	20
1,400	1,105	210	56	15

Table 14.2 Average values for grain size and grain height of 8YSZ-MOD layers (MOD 55–MOD 62) for different temperature treatments

Temperature (°C)	Grain size (nm)	Height (nm)
500	Amorph.	–
650	Amorph.	–
850	28	17
1,250	330	85
1,350	560	103
1,400	690	121
1,700	Inhomog.	–

Below 650°C and at 1,700°C annealing temperature no grains could be observed. Higher annealing temperatures result in increasing grain sizes. At 1,700°C no grains could be observed but an inhomogeneous surface morphology instead

14.2 AFM Characterization and Grain Size Analysis

The investigated nanocrystalline layers were made of yttria-stabilized zirconia (YSZ), in this case zirconium oxide ZrO_2 doped with 8 mol% yttrium oxide Y_2O_3 (8YSZ). The layers were produced on a substrate of sapphire and 8YSZ, at temperatures between 500 and 1,700°C. The analysis of the topography, shown in Fig. 14.1, was made with the AFM in contact mode and shows the grain size dependence of the temperature of the last production step.

The grain size distribution of the different samples was analyzed manually and automatically by software. The results were compared with XRD measurements (Fig. 14.2), which were made by the Institute of Materials for Electrical Engineering (IWE). The analysis of the AFM data was reliable for grain sizes above 20 nm, but became somewhat inaccurate below. This is due to tip convolution artifacts, because in this range the grain diameter is of the same order as the tip radius. For smaller grain sizes, the AFM analysis yields slightly higher values, whereas for bigger diameters the AFM and XRD data coincide.

In summary, a strong correlation between grain size distribution and annealing temperature could be observed. The higher the temperature, the bigger the grains. This is in good correspondence with the results achieved with XRD. Additionally, the height of the grains could be measured.

14.3 Chemical Contrast Imaging

In order to differentiate chemical inhomogeneities at sample surfaces on the nanometer scale, the method of CCI as a new dynamic AFM mode for the measurement of friction forces between the AFM tip and a sample surface was developed in our workgroup and was patented in 2006 [4]. The new method was employed to study the chemical homogeneity of the LSC50-MOD layers, and the mode was also

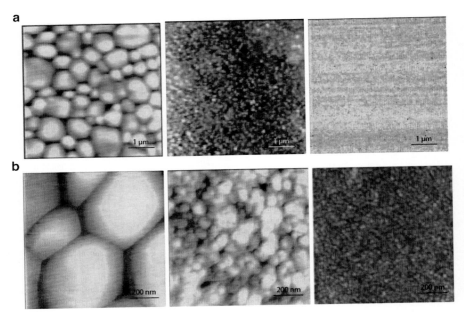

Fig. 14.1 (a) Topography analysis of 8YSZ-MOD layers produced at various temperatures: *left* 1,400°C, *middle* 1,000°C, *right* 650°C. Scan size 5 × 5 μm². (b) Topography analysis of 8YSZ-MOD layers produced at various temperatures: *left* 1,400°C, *middle* 1,000°C, *right* 650°C. Scan size 1 × 1 μm²

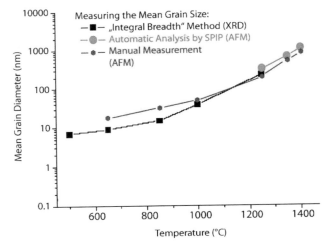

Fig. 14.2 Grain size analysis: Black curve: XRD measurements by the Institute of Materials for Electrical Engineering (IWE). *Red curve*: manual analysis; *blue curve*: automatic analysis of AFM images. The grain size of the 8YSZ-MOD layers depends on the temperature of the last annealing step of the production process. The higher the temperature, the bigger the grains become, from 7 nm at 500°C up to 1,050 nm at 1,400°C

methodically improved for further proposes. Hence, the principles of the method should be illustrated shortly in the next section.

As in conventional force modulated microscopy (FMM) during CCI, the AFM tip is in contact with the sample. Also the distance between sample and the AFM cantilever is modulated sinusoidally. In the case of CCI, this is done by moving the z-position of the sample with a higher frequency compared with the imaging scan frequency. As in FMM, the amplitude of the first harmonic oscillation of the normal force signal (FMM amplitude) is recorded with lock-in technique. In contrast to conventional FMM, soft cantilevers with bending force constants small with respect to the stiffness of the contact between AFM tip and sample surface are used. Thus, the recorded signal is less sensitive to the elastic properties of the contact. The parameters are chosen so that the measurement is rather sensitive to the friction forces occurring within the contact.

Typically, the cantilever is mounted at an angle of 15° with respect to the sample surface. Thus, the AFM tip in contact with the sample surface is forced to follow the vertical modulation imposing also a lateral movement of the tip along the cantilever axis. Friction forces acting against such a lateral movement cause buckling of the cantilever beam, which also contributes to the normal-force signal and therefore can be measured within the FMM amplitude signal.

Using appropriate cantilevers combined with the selection of appropriate parameters especially the modulation amplitude, this signal is very sensitive detecting the friction forces between the AFM tip and the sample. Therefore, it can be used to distinguish chemically different areas on the surface in case these induce different tribological properties on the surface.

In comparison with the most-commonly used friction-sensitive AFM mode, the lateral force microscopy, CCI has the following advantages: The use of a lock-in technique results in a better signal stability and an improved S/N ratio, and the CCI is less affected by topographic influences. Furthermore, the method is easy to implement to existing setups.

The examination of the chemical homogeneity of the MOD layers is an important issue. Experimental data indicated a phase separation within the LSC50-MOD layers. In case the different phases have surfaces with different tribological properties, this could be detected by CCI. Then not only the presence of chemical inhomogeneities could be detected, but also the local distribution of the different phases. CCI measurements of LSC50-MOD samples did not give clear evidence for chemical inhomogeneities. However, the marked contrast, as shown in Fig. 14.3, could possibly indicate a phase contrast. Furthermore, not all grains of the sample may have contact to the sample surface and as the AFM only detects chemical contrast at the surface, they cannot be seen with this method.

Fig. 14.3 Chemical contrast imaging of a LSC50-MOD layer ($3\mu m \times 3\mu m$). The marked areas in the CCI image indicate a phase contrast

14.4 Electrostatic Force Microscopy (EFM)

Since the development of scanning force microscopy in the eighties, various techniques have been proposed, based on different interactions between the probe and the sample. The long range of electrostatic interactions makes them especially suitable for noncontact imaging of conducting and insulating surfaces. By applying a bias voltage between a conducting force microscopy tip and the sample, electrostatic force microscopy (EFM) can be used to measure local variations of capacitance, surface potential, charge or dopant distribution, topography, and dielectric properties of metallic and insulating surfaces [8].

The potential difference between the tip of the EFM and the sample, $\Phi_t - \Phi_s$, depends on the contact potential, U_{cp}, and the external bias voltage, U_{bias} (see Fig. 14.4) [9]. Two kinds of charges contribute to the electrostatic interaction: free charges, induced by the contact potential, U_{cp}, and the external bias voltage, and localized charges on the tip, q_t, and on the sample, q_s. We assume that there is no oxide layer on the surface of the tip and therefore set $q_t = 0$.

Now, three components contribute to the force between tip and sample:
First, the Coulomb force between the localized charges.

$$F_1 = \frac{1}{4\pi\varepsilon_0 z^2}(q_s - q_t)^2, \quad \text{for } q_t = 0 \rightarrow F_1 = \frac{1}{4\pi\varepsilon_0 z^2}q_s^2$$

The value z is the distance between tip and sample.

Second, the Coulomb force between the localized and the induced charges, depending on the potentials U_{cp} and U_{bias}.

$$F_2 = \frac{C}{4\pi\varepsilon_0 z^2}(U_{bias} - U_{cp})q_s \quad (\text{for } q_t = 0)$$

The value C is the capacity of the tip–sample-arrangement.

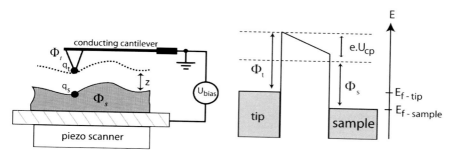

Fig. 14.4 (*Left*) Principle of the EFM, (*right*) energy diagram. The cantilever of the EFM is being bent by the Coulomb force between tip and sample. The force depends on the localized charges q_s on the sample and q_t on the tip, the contact potential U_{cp}, and the external bias voltage between tip and sample U_{bias}. The contact potential U_{cp} results from the different work functions of the materials of the tip and the sample

The third component is the Coulomb force between the free charges on the sample [10]:

$$F_3 = \frac{1}{2}\frac{dC}{dz}\left(U_{bias} - U_{cp}\right)^2.$$

The sum of these components describes the complete electrostatic interaction, when an external potential U_{bias} is applied.

$$F_{elec} = \frac{1}{4\pi\varepsilon_0 z^2}q_s^2 + \frac{C}{4\pi\varepsilon_0 z^2}\left(U_{bias} - U_{cp}\right)q_s + \frac{1}{2}\frac{dC}{dz}\left(U_{bias} - U_{cp}\right)^2$$

If we assume that there are also no localized charges on the sample ($q_s = 0$), the equation reduces to:

$$F_{elec} = \frac{1}{2}\frac{dC}{dz}\left(U_{bias} - U_{cp}\right)^2 \quad (\text{for } q_s = 0).$$

In the EFM, the cantilever is now excited by a small piezoceramic to oscillate at its mechanical resonance frequency, while the bias voltage is applied between the tip and sample. The electrostatic force on the cantilever depends on the height, z, of the tip above the sample. The oscillation causes the tip to move through a nonlinear electrical potential, which alters the effective resonance frequency of the system. The local force gradient dF/dz, caused by the potential difference $\Phi_t - \Phi_s$ can now be detected by measuring the change of the amplitude of the oscillating cantilever or by measuring the phase shift between the actual oscillation of the cantilever and the driving oscillation (see Fig. 14.5).

The frequency and phase shift can be described as:

$$\Delta\omega = -\frac{\omega_0}{2k}\frac{dF}{dz} \quad \text{and} \quad \Delta\phi = -\arcsin\left(\frac{Q}{k}\frac{dF}{dz}\right)$$

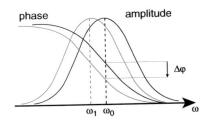

Fig. 14.5 Phase and amplitude shift in the EFM. The electrostatic force shifts the resonance frequency of the oscillating cantilever from ω_0 to ω_1. This causes a reduction of the amplitude and a phase shift at the original resonance frequency ω_0

With the derivation dF_{elec}/dz

$$\frac{dF}{dz} = \frac{1}{2}\frac{d^2C}{dz^2}(U_{bias} - U_{cp})^2$$

the phase shift can be calculated as a function of U_{bias} and U_{cp}.

$$\Delta\phi = -\arcsin\left(\frac{Q}{2k}\frac{d^2C}{dz^2}(U_{bias} - U_{cp})^2\right),$$

where Q is the quality factor and k the spring constant of the cantilever.

To eliminate the influence of the sample topography, the electric interaction is measured in a certain distance above the surface in the noncontact mode. To obtain this fixed height, the sample topography profile along a scan line is determined in a first scan in intermittent contact mode. Then in a second scan, the oscillating tip is moved in a constant height of a few ten nanometers above the sample and the change of amplitude and the phase shift are recorded. This is described as lift mode (Fig. 14.6).

The EFM delivers very good qualitative results. However, a quantitative analysis of the results is very difficult. The method is still in development and no exact physical model exists by now, which allows a quantitative analysis of the data.

14.5 Implementation and Test of the EFM Method

To convert the existing AFM into an EFM, conducting connections to the tip and the sample were needed. Connecting the tip required special cantilevers, which are coated with a hard conducting material, such as titanium, platinum, or tungsten carbide. The cantilever mount and the holding clamp had to be insulated from the AFM-head and a contact to the cantilever had to be established by connecting a lead to the holding clamp.

For connecting the samples, special sample holders were designed, consisting of standard magnetic pucks, on witch up to four electrically insulated terminals were mounted. From these terminals, very thin highly flexible wires lead to an external

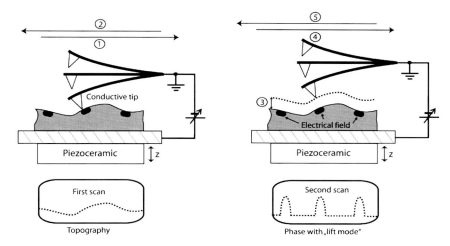

Fig. 14.6 Lift mode measurement in the EFM. First the topography profile of the sample is measured in a first scan in intermittent contact mode (1). Then the cantilever is positioned at the start of the same scan line (2) in a certain height of a few ten nanometers above the surface (3). The second scan follows the topography of the sample in this constant height, while recording the phase shift of the oscillation of the cantilever (4). Then the tip is positioned at the start of the next scan line (5)

voltage source or other measuring equipment. To connect the sample surface, thin but rigid Nickel wires lead from the terminals to the surface. The wires were bent so that they would press their free end on to the desired area of the sample and provide an electrical contact. In case of too high a contact resistance, the conductivity could be improved by evaporating aluminum electrodes onto the sample surface and placing a small portion of conductive silver to the contact area.

Bias voltages of up to 7.5 V were applied between the tip and the sample and lift heights of 30–50 nm were selected. The method was tested on a structure of intercalated gold leads on a silicon substrate and on indium tin oxide clusters, also deposited on silicon (Fig. 14.7).

As a further lifelike test sample metallic nanowires were developed, characterized with other methods like SEM and scanning Auger spectroscopy [11], and then investigated by EFM. These wires have the prospect of being used for future local contacting of individual grains and grain boundaries. A scheme of the wire profile is shown in Fig. 14.8: Cu wires lie next to each other, separated by spacers of nonconductive CuO. Additionally, CuO particles are randomly distributed on the surface.

The AFM topography measurement depicted in Fig. 14.9 shows clearly the raised, about 30 nm wide Cu wires and the lower spacers with a width of about 15 nm. Additionally round islands lying on the surface can be seen. From the EFM phase image next to it, one can see that the higher topographic lines are conductive, i.e., consist of copper. In contrast, the spacers, just like the round islands, appear nonconductive, both therefore confirmed to consist of CuO.

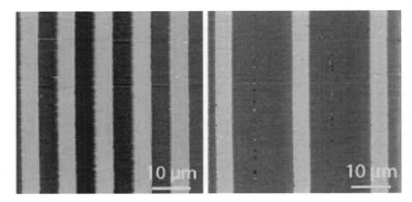

Fig. 14.7 Test of the EFM on a structure of intercalated gold leads on a silicon sample. Left: First scan of the topography. Right: Second scan of the phase shift at the same position, showing the electrical interaction between tip and sample. Here only every other line is detectable. These were connected to a bias voltage of 3 V, while the others were not connected and therefore do not show any contrast. In order to get reproducible results, at the beginning of each measuring session the EFM was calibrated by scanning this sample

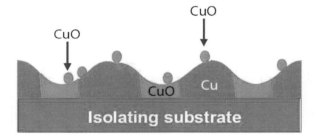

Fig. 14.8 Scheme of the profile of a copper nanowire sample confirmed by SEM and scanning Auger measurements. Cu wires lie next to each other, separated by spacers consisting of CuO [11]

Figure 14.10 proves the very high resolution that can be achieved with our EFM system. Indium tin oxide clusters were deposited on a silicon surface (Sample courtesy of the group of Prof. Hahn, INT). The smallest visible features visible are individual clusters with feature sizes down to 10 nm.

14.6 Electrical Characterization of 8YSZ-MOD Layers

After the investigation of the dependence of the grain size from the production temperature, the aim of the following experiment is to study the influence of the grain sizes on the conductivity of the layers. The grains and the grain boundaries may play different roles for the conductivity.

14 Studying Functional Electrode Structures   155

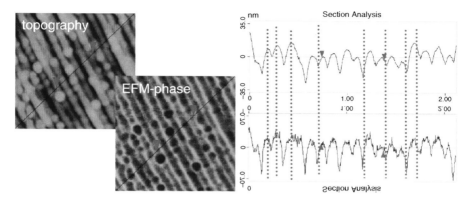

Fig. 14.9 Comparison of the topography (*left*) and the corresponding EFM phase signal (next to left, bias voltage 5 V) of a alternating Cu/CuO nanowire structure. The higher topographic lines are corresponding to the areas of higher conductivity, whereas the valleys have a lower conductivity. Image size 1.7 μm × 1.7 μm. The two cross sections (*right*) are taken along the lines drawn in the topography and EFM image, respectively

Fig. 14.10 Test of the EFM on indium tin oxide clusters deposited on a silicon surface at a bias voltage of 3 V. The smallest features visible in this image (*red circles*) are individual clusters with feature sizes of 30 down to 10 nm. This proves the very high resolution that can be obtained with our EFM setup

To investigate the conductivity in the nanoscale regime, the layers were now imaged with the EFM. The contact wires were connected directly to the sample surface with a drop of conductive silver. The cantilever was coated with platinum/iridium and kept on ground potential, while the bias voltage was applied to the sample. The lift mode height was set to 50 nm.

In the following EFM images, a bright contrast corresponds to a higher electronic conductivity on the sample.

The EFM investigation of different 8YSZ-MOD layers shows that the layers are macroscopically conductive. Figure 14.11 shows an EFM image of a layer produced at 1,400°C. It shows a brighter contrast at the grain boundaries than in the grains

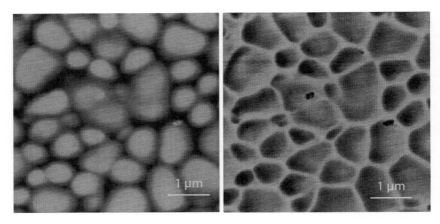

Fig. 14.11 *Left*: topographic AFM image of an 8YSZ-MOD layer, recorded in intermittent contact mode. The layer was prepared at a temperature of 1,400°C. *Right*: EFM phase shift image of the same area, recorded in lift mode with a bias voltage of 7.5 V. It is evident, that the grain boundaries show a brighter contrast than the grains themselves, indicating a higher conductivity

themselves. It could also be shown that the contrast is proportional to the bias voltage. This contrast shows a higher conductivity at the grain boundaries.

This result is surprising so far, as there should be no known ionic conductivity of such layers at room temperature. Ionic conductivity begins to appear at temperatures above 200°C. This leads to the conclusion that there is electronic conductivity, which cannot be the case for this material.

A possible explanation for this effect, described in Fig. 14.12 is the water layer, which is present on every surface at ambient conditions, at which these measurements were made. The water will aggregate preferentially in the deeper grain boundaries. The water molecules can easily be polarized by the changing electric field, caused by the oscillating cantilever tip. Thus, mirror charges are induced, which alter the interaction between tip and sample, resulting in the observed contrast.

Another method for electrical characterization is the conducting AFM (C-AFM). Here, the tip is brought in direct contact with the sample surface and the current caused by the bias voltage is measured. This method was also implemented, but in case of the 8YSZ-MOD layers, no conductivity could be observed (Fig. 14.13). This is expected for this material, because The Material itself is an oxide ceramics, which prevents a direct conducting contact for electrons. There should be also no electric conductance through the water layer, because taking into account the hydrophilic behaviour of 8YSZ-MOD and the roughness of the sample, there will be no water layer which is thick enough to show such a conductivity, which could produce such EFM signals. And there has been also no electric contact to the surface of the sample. It was connected to the EFM electrodes through the sample plate.

The comparison of the AC measurement by EFM with the DC measurement by C-AFM indicates that there is no ionic or electronic conductivity at room temperature. The measured AC electronic conductivity stems from the polarization of the water film always present on surfaces at ambient conditions.

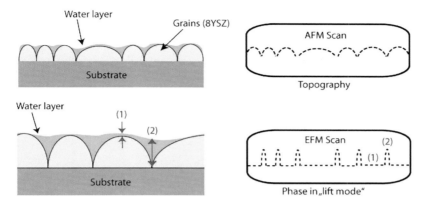

Fig. 14.12 *Left*: The ambient air humidity creates a water layer on the sample surface. Due to the topography and capillary effects the layer varies in thickness between the grains (1) and the grain boundaries (2). *Right*: an increased thickness of the water layer causes a higher contrast in the EFM measurement

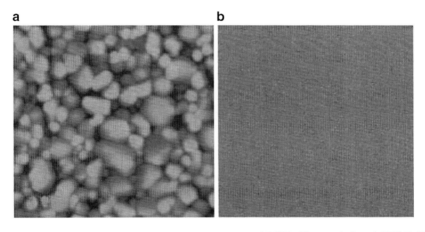

Fig. 14.13 C-AFM measurement at room temperature of MOD 59 annealed at 1,000°C. The image size is 3 μm × 3 μm. (**a**) Topography showing the expected roughness of the surface. (**b**) The C-AFM measurement showed no contrast within the accuracy of the measurement, i.e. there is no evidence for electronic conductivity (see text)

14.7 Summary

In order to be able to perform the electronic measurements, an AFM system was adapted and extended for EFM and C-AFM. To get reproducible results, a special calibration sample has been designed and built. Furthermore, a more lifelike sample consisting of CuO covered Cu nanowires separated CuO spacers was developed and analyzed by SEM and scanning Auger spectroscopy. The EFM results were consistent with these measurements proving the accuracy of our instrument. The

EFM measurements with the extended system showed extraordinarily fine features in the 10-nm regime. To the best of our knowledge, this has not been observed under ambient conditions to date in the literature.

EFM measurements on 8YSZ-MOD layers showed a clear contrast between crystallites and grain boundaries in which the crystallites appear less conductive than the grain boundaries. As there should be no electronic conductance at all at room temperature within the 8YSZ-MOD layers, further examinations were carried out. A water layer with a different thickness on the crystallites compared with the grain boundaries could be proposed as producing this effect. With C-AFM measurements, no evidence for the conductivity at room temperature of the MOD layers could be observed.

Acknowledgments We thank E. Ivers-Tiffée and C. Peters from the Institute of Materials for Electrical Engineering for supplying the fuel cell samples and also acknowledge the EFM test samples from Gabi Schierning and Norman Mechau from the Institute of Nanotechnology.

This work was supported by the Deutsche Forschungsgemeinschaft within the DFG-Center for Functional Nanostructures (CFN) and by the Landesstiftung Baden-Württemberg within the Research Network "Functional Nanostructures".

References

1. Z. Shao, S.M. Haile, Nature **431**, 170 (2004)
2. R. Krüger, A. Weber, E. Ivers-Tiffée, C. Peters, M. Bockmeyer, Mater. Res. Soc. **928** (2006)
3. H.L. Tuller, Solid State Ionics **131**, 143 (2000)
4. M. Müller, Th. Schimmel, Method and device for determining material properties, PCT Patent application, WO2006/097800 A2, 21.09.2006
5. A. Pfrang, M. Müller, Th. Schimmel, Chemical contrast imaging, *Photonik* 02/08 (2008)
6. Th. Schimmel, R. Gröger, M. Müller, H. Gliemann, Chemical contrast imaging and the study of nanoparticle adhesion on surfaces with the tip of an atomic force microscope. Invited Talk, Workshop "Engineered Nanoparticles in the Aquatic Environment", Ladenburg, Germany, June 30, 2008
7. Th. Schimmel, Chemical patterning and chemical contrast imaging with the tip of an AFM. Invited Talk, Materials Research Conference MR-2007, Saarbrücken, Germany, September 03–07, 2007
8. C.C. Williams, Y. Martin, H.K. Wickramasinghe, J. Appl. Phys. **61**(10), 4723 (1987)
9. M. Elliot, C.H. Lei, A. Das, J. Macdonald, Appl. Phys. Lett. E **83**, 482 (2003)
10. P. Girard, Nanotechnology **12**, 485 (2001)
11. S. Zhong, P. Dupeyrat, R. Gröger, M. Wang, Th. Koch, Th. Schimmel: *Periodical Nanostructured "Multiline" Copper Films Self-Organized by Electrodeposition: Structure and Properties*. J. Nanosci. Nanotechnol., in press 2009

Part IV
Nanobiology

Chapter 15
Integrated Lab-on-a-Chip System in Life Sciences

S. Thalhammer, M.F. Schneider, and A. Wixforth

Abstract Surface acoustic waves are employed to efficiently actuate and manipulate smallest amounts of fluids on a chip. The interaction between the surface wave and the fluid on a chip leads to acoutic streaming within the fluid, which can be used to pump and mix within a closed volume. At somewhat higher surface wave amplitude, small fluid volumes like droplets can be actuated as a whole. Our technique yields a very versatile approach toward a programmable fluidic microprocessor for which we give a few representative application examples, including a complete micro total analysis system with polymerase chain reaction on a chip.

15.1 Lab-on-a-Chip Systems

About a decade ago, a handful of researchers began discussing an intriguing idea. Could the equipment needed for everyday chemistry and biology procedures be shrunk to fit on a chip in the size of a fingernail? Miniature devices for, say, analyzing DNA and proteins should be faster and cheaper than conventional versions. Lab-on-a-chip is an advanced technology that integrates a microfluidic system on a microscale chip device. The "laboratory" is created by means of channels, mixers, reservoirs, diffusion chambers, integrated electrodes, pumps, valves, and more. With lab-on-a-chip technology, complete laboratories on a square centimeter can be created.

The goal of the lab-on-a-chip technology is to automate standard laboratory processes and to conduct chemical and biochemical analysis in a miniaturized format. This makes it fast and cost efficient with small reagent consumption and less waste generation; the results of a research can be obtained within a few seconds instead of hours or days.

Microfluidics deals with the handling and manipulation of minute amounts of fluids: volumes thousands of times smaller than a common droplet. Microfluidics means measuring in microliters, nanoliters, or even picoliters. The microfluidics field lies at the interfaces between biotechnology, medical industry, chemistry, and microelectromechanical systems (MEMS). During the past 20 years, microfluidics, micrometer-scale total analysis systems (μTAS) or so-called "lab-on-a-chip" devices have revised interest in the scaling laws and dimensionless groups for downscaling purposes [1]. Microfluidics already started twenty years ago, principally in inkjet printer manufacturing. The mechanism behind these printers is based on microfluidics; it involves very small tubes carrying the ink for printing. The tubes can combine and isolate from each other to change the tone of the colors as they appear on the page. The same etching techniques used to make semiconductor chips can create channels, tubes, and chambers in silicon or glass substrates, which can be layered on top of one another to make more complex 2-D and 3-D structures. Chip designs can also be stamped or cut into plastic sheets.

While the earliest reported microscale devices consisted of channels etched in hard substrates such as silicon [2, 3], glass [4] and plastic [5], MEMS fabrication technologies have been increasingly applied to fabricate highly sophisticated devices from a variety of materials, including soft elastomers such as polydimethylsioxane (PDMS) [6] with hundreds of microchannels and integrated sensors to measure physiological parameters. It is the next step in laboratory automation, resulting in:

- A decrease in reagent consumption and waste
- A reduction of cost per analysis
- Faster analyses and results in a few seconds
- Safer chemical experiments and reactions
- An improved data quality
- Better controllable process parameters in chemical reactions
- An increased resolution of separations

The ability to precisely control parameters such as substrate, flow rate, buffer composition, and surface chemistry in these microscale devices makes them ideal for a broad spectrum of cell-biology-based applications. It ranges from high-throughput screening of single cells and 3-D scaffolds for tissue engineering to complex biochemical reactions like polymerase chain reaction (PCR) and drug detection. Microscale devices offer the possibility of solving system integration issues for cell biology, while minimizing the necessity for external control hardware. Many applications, such as single gene library screening, are currently carried out as a series of multiple, labor-intensive steps required in the array process, e.g., DNA amplification, reporter molecule labeling, and hybridization. While the industrial approach to complexity has been to develop elaborate mechanical high-throughput workstations, this technology comes at a price, requiring considerable expenses, space, and labor in the form of operator training and maintenance. For small laboratories or research institutions, this technology is simply out of reach. Devices consisting of addressable microscale fluidic networks can dramatically simplify the screening process, providing a compartmentalized platform for nanoliter aliquots. However, making miniature labs is not just a question of scaling down conventional equipment. Nanoliter volumes of liquid behave in curious ways. For example, we are used to seeing liquids mix by turbulent flow, as illustrated by the way cream swirls into coffee, but such turbulence does not occur in closed channel or tubing systems just a few micrometers wide. Streams of liquid flow alongside each other over short distances without mixing. At the other end of the application spectrum, the ability to regulate fluid flow within open microscale devices with nanoliter precision has also generated interest in using them as tools for tissue engineering or bionic applications. Advances in substrate micropatterning techniques to mimic capillaries (e.g., bionic approaches), the implementation of biocompatible and biodegradable substrates, and the ability to pattern surfaces with molecules to simulate cell adhesion have provided researchers with excellent tools to understand complex biophysical processes, e.g., the adhesion of von Willebrand factor (VWF) fibers to stop bleeding under high shear-stress conditions as found in small blood vessels [7].

15.2 General Manipulation of Cells and Cell Components in Microdevices

Recent advances in the field of genomics and proteomics have catalyzed a strong interest in the understanding of molecular processes at the cellular level like gene expression and cell biomechanics. In combination, this leads to the novel transdisciplinary field of systems biology. Microscale devices are being applied to both the manipulation and the interrogation of single cells in nanoliter volumes. At the single cell level immobilization, sorting and assay approaches are briefly summarized (see Table 15.1). The ability to manipulate single cells and cell components has been accomplished over the past few decades using a variety of techniques, including mechanical approaches, e.g., micropipettes and robotic micromanipulators [8], optical tweezers [9, 10], and microelectrodes [11, 12]. Adapting these general classes of techniques to nanofluidic systems, using microchannels or microfabricated structures incorporating nanoliters of fluid, offers an unprecedented level of control over positioning, handling and patterning of single cells.

Table 15.1 Microdevice approaches to immobilize, sort, monitor and culture single cells

Immobilization of cells	Mechanical cell trapping
Cell sorting	Biomolecular cell trapping
	Electrokinetic cell trapping
	Electroosmotic and dielectrophoretic cell sorting
	Pressure based cell sorting
	Optical sorting
Monitor cell physiology	Single-cell electroporation
	Cellular nucleic acid isolation
	Nucleic acid amplification/detection
	Nanofluidic electrophoresis
	Proteomics applications
	Quantifying intracellular protein and small molecule detection
Cell culture in microdevices	Surface chemistry and cell culture
	Microchannel geometry and cell culture
	Low control in microscale culture systems

Several groups have recently reported on the development of nanofluidic systems to mechanically manipulate and isolate single cells or small groups of cells in microscale tubing and culture systems. The Quake group used multilayer soft lithography, a technology to create stacked 2-D microscale channel networks from elastomers, to fabricate integrated PDMS-based devices for programmable cell-based assays [13]. They applied the microdevice for the isolation of single *Escherichia coli* bacteria in subnanoliter chambers and assayed them for cytochrome c peroxidase activity. Khademhosseini and coworkers reported on the use of polyethylene glycol (PEG)-based microwells within microchannels to dock small groups of cell in predefined locations. The cells remained viable in the array format and were stained for cell surface receptors by sequential flow of antibodies and secondary fluorescent probes [14]. The trapping of cells using biomolecules in nanofluidic systems has been demonstrated using antibodies and proteins with high affinity to the target cell (for review see [15]). Chang et al. used square silicon micropillars in a channel coated with the target protein, an E-selectin-IgC chimera, to mimic the rolling and tethering behavior of leukocyte recruitment to blood vessel walls [16]. Using electric fields both to induce flow and to separate molecules is widely adapted to microscale devices to separate nucleic acids and proteins [17, 18]. For cell-capture dielectrophoresis, in which a nonuniform alternating current is applied to separate cells on the basis of their polarizability, has been adapted to microscale devices [19].

In an effort to miniaturize the cell-sorting process, microscale devices have been fabricated that use different strategies, including electrokinetic, pressure and optical deflection-based methods [20]. While the first prototypes developed in the late 1990s were capable of tens of cells per second, state-of-the-art microscale fluorescent-activated cell sorters are quite comparable to conventional sorters [21]. Wang and coworkers implemented a fluorescence-activated microfluidic cell sorter and evaluated its performance on live, stably transfected HeLa cells expressing a fused histone-green fluorescent protein [22]. Electroosmotic flow, in which flow is induced by applying a voltage potential that induces flow through the migration of ions in solution, was one of the earliest technologies to be used in microscale flow cytometers. Fu and coworkers developed a simple PDMS T-microchannel sorter device consisting of a single input and two electrode connected collection outputs [20]. The sorting of cells was accomplished by switching the applied voltage potential between the two outputs. Flow switching in pressure-based cell sorting, to the collection and waste outputs, has been achieved using by integrated elastomeric valves in multilayer PDMS-based devices and hydrodynamic switches [21]. Ozkan et al. used vertical cavity surface-emitting laser arrays to trap and manipulate individual cells in PDMS microchannels [23]. While field-trapping strengths are characteristically weak for larger objects like mammalian cells, the use of photonic pressure to repel cells has been shown to be an effective tool for cell sorting microfabricated devices. Besides multicellular sorting, single cell analysis plays an important role in molecular diagnosis. All basic microdissection strategies of cellular nucleic acid isolation ranging from single cells to cell fragments are reflected in the following chapter.

It is not the intention to provide an entire comprehensive list of microdevices to monitor single-cell physiology. A list of the major fields of research is summarized in Table 15.1 and the present status and novel achievements in the monitoring cell physiology field are reviewed (for history, theory, and technology see [24]; for analytical standard operations and applications see [25]; for latest trends see [26]). Cellular nucleic acid isolation and nucleic acid amplification and detection are relevant to the developed novel lab-on-a-chip system are described in detail subsequently.

15.3 Actuation of Lab-on-a-Chip Systems

Micropumps are one of the components in three-dimensional Lab-on-a-Chip systems with a large variety of operating principles; they can be divided in two mean categories: mechanical pumps and nonmechanical pumps (for review see [27]). Mechanical pumps usually utilize moving parts such as check valves, oscillating membranes, or turbines for delivering a constant fluid volume in each pump cycle [28]. These are mainly used in macroscale pumps and micropumps with relatively large size and large flow rates. Since the viscous force in microchannels increases in the second order with the miniaturization, mechanical pumps cannot deliver enough power in order to overcome its high fluidic impedance. Nonmechanical pumps add momentum to the fluid for pumping effect by converting another energy form into the kinetic energy and therefore have the advantages in the microscale. For flow rates larger than 10 ml/min, miniature pumps or macroscale pumps are the most common solution. The typical operation range of positive displacement micropumps lies between 10 and several mircroliters per minute. For flow rates less than $10\,\mu l/min$, alternative dynamic pumps are needed for accurate control of these small fluid amounts. With these flow rates, most of the pumps are working in the range of Reynolds number from 1 to 100, and therefore in a laminar regime.

Another mechanical pump type, the rotary pump, can be realized with micro machining techniques for pumping highly viscous fluids. Ahn and Allen presented a micropump with a microturbine as the rotor in an integrated electromagnetic motor. The high aspect ratio structures with $160\,\mu m$ chamber height were fabricated using photolithography of polyimide. They achieved a flow rate of $24\,\mu l/min$ [29]. A two gear wheel rotary micropump with a flow rate of $55\,\mu l/min$ in $500\,\mu m$ chamber heights was presented by Doepper and coworkers. The gears forced the fluid along by squeezing it to an outlet [30]. Centrifugal microfluidic platforms are of particular interest for assay integration as their artificial gravity field intrinsically implements a pumping force as well as an established method for particle separation without actuation apart from a standard rotary drive. The Zengerle group presented a plasma extraction method from whole blood on a rotating disk with a capillary system for subsequent on-disk processing [31].

Ultrasonic pumping is a gentle pump principle with no moving parts, heat, and strong electric field involved. Yang and coworkers designed an active micromixer based on ultrasonic vibrations and successfully tested their system using water and uranine [32]. The pump effect is caused by the acoustic streaming, which is induced by a mechanical traveling wave. The mechanical wave can be a flexural plate wave [33, 34] or a surface acoustic wave (SAW) [35, 36]. The mechanical waves are excited by interdigital transducers placed on a thin membrane coated with piezoelectric film [33, 34] or on a piezoelectric bulk material [35, 36]. This is an alternative to the later-described electrowetting-based transport of droplets.

All mechanical pumps require a mechanical actuator, which generally converts electric energy into mechanical work. They can be divided into external and integrated actuators (for review see [37]).

External actuators include electromagnetic actuators with solenoid plunger and external electric field, disk type or cantilever type piezoelectric actuators, stack type piezoelectric actuators, pneumatic actuators, and shape memory actuators. The biggest drawback of external actuators is their large size, which restricts the size of the whole micropumps. The advantage is the relatively large force and displacement generated by external actuators.

Integrated actuators are electrostatic actuators, thermopneumatic actuators, electromagnetic actuators, and bimetallic actuators. Despite their fast response time and good reliability, electrostatic actuators cause small force and very small stroke. With special curved electrodes, electrostatic actuators are suitable for designing micropumps with a very low power consumption. Thermopneumatic actuators generate large pressure and relatively large stroke. This actuator type was therefore often used for mechanical pumps. Thermopneumatic actuators and bimetallic actuators require a large amount of thermal energy for their operation, and consequently, consume a lot of electric power. High temperature and complicated thermal management are further disadvantages of these kinds of actuators. Electromagnetic actuators require an external magnetic field, which also restricts the pump size. Their large electric current causes thermal problems and high electric energy consumption (for review see [37]).

Nonmechanical pumps can be divided into electrohydrodynamic, electrokinetic, phase transfer, electrowetting, electrochemical, and magentohydrodynamic pumps (for review see [27, 38]). Electrokinetic pumping and particle manipulation principles are based on surface forces and thus gain impact within the microdimensions due to the increased surface-to-volume ratio (SVR). This advantage combined with the simple setup of electrokinetic systems, which basically consist of microfluidic channels and electrodes, was used in lab-on-a-chip applications for the analysis of chemical compounds. The electrohydrodynamic induction pump (EHD) is based on the induced charge at the material interface. A traveling wave of electric field drags and pulls the induced charge along the wave direction. It was first presented by Bart et al. and with similar designs by Fuhr and coworkers [39, 40]. A fluid velocity of several hundred microns per second can be achieved with this pump type. In contrast to the EHD-pumps, electrokinetic pumps utilize the electric field for pumping conductive fluid. The electrokinetic phenomena can be divided into electrophoresis and electro-osmosis. Electrophoresis is the effect by which charged species in a fluid are moved by an electrical field relative to the fluid molecules. Electrophoresis is used to separate molecules like DNA molecules depending on the size. In contrast to electrophoresis, electro-osmosis is the pumping effect of a fluid in a channel under the application of an electrical field. A surface charge exists on the channel walls and the electro-osmotic effect is used for pumping fluid in small channels without applying a high external pressure. The most common application of electro-osmotic pumps is the separation of large molecules like DNA or proteins. Harrison et al. presented a system that generates a fluid velocity of 100 μm/s with a field strength of 150 V/cm [4]. Gel electrophoresis for separating DNA molecules in microchannels with field strengths ranging from 5 to 10 V/cm was shown by Webster and coworkers [41]. In order to overcome the high fluidic impedance caused by viscous forces in small channels, phase transfer pumps can be an alternative. This pumping principle uses the pressure gradient between the gas and liquid phase of the same fluid for pumping it. The first phase transfer pump was presented by Tagagi et al., where the alternate phase change is generated by an array of ten integrated heaters [42]. A much smaller pump based on surface micromachining with six integrated polysilicon heaters in a channel with 2 μm height and 30 μm width was described by Jun and Kim [43]. With this pump type, a flow velocity of 160 μl/s or flow rates less than one nanoliter per minute can be achieved. The electro-wetting pump was suggested by Matsumoto and Colgate, where the dependence of the tension between solid–liquid interface on the charge of the surface is used for actuation. Lee and Kim reported a micro actuator based on the electro-wetting of mercury drops, which can be used for driving a mechanical pump with check valves [44]. A micro liquid dosing system is presented, which allows bidirectional manipulation of fluids (i.e., pushing out and pulling in of liquids) by the electrochemical generation and removal of gas bubbles. Bidirectionality is obtained by reversal of the actuation current thereby causing the earlier produced gasses to react back to water. This reduction of gas volume actively pulls liquid back into the system. The electrochemical actuator electrodes have been specially designed to perform the simultaneous measurement of conductivity, via which the total amount of gas can be estimated [45]. Magnetohydrodynamic pumps use the Lorentz force acting on a conducting solution to propel an electrolytic solution along a microchannel etched in silicon. The micropump described by Lemoff and Lee has no moving parts, produces a continuous (not pulsatile) flow, and is compatible with solutions containing biological specimens [46]. The use of magnetic liquids is an innovative actuation method for liquid handling, since ferrofluids have shown their great potential [47]. Yamahata et al. presented a ferrofluid micropump where the magnetic plug is externally actuated by a motorized permanent magnet. Water has been successfully pumped at a flow rate of 30 μl/min at backpressures of up to 25 mbar. [48]. Transportation of aqueous droplets containing hydrophilic magnetic beads in a flat-bottomed, tray-type reactor filled with silicon oil and actuated by an external magnetic force was presented by Ohashi et al. [49]. They performed PCR reactions in 3 μl virtual reaction chamber (VRC) droplets in 11 min with 30 cycles resulting in PCR products ranging from 12^6 to 12^{19} basepairs.

15.4 Lab-on-a-Chip Concepts

The need for faster and cheaper technologies to extract biological information, both at the molecular and cellular levels, has driven the trend to miniaturize laboratory techniques in the last two decades. Just as integrated circuits revolutionized information technology, multiplexed microscale devices capable of manipulating and processing DNA and proteins in nanoliters of fluid have the potential to have the same impact on biology and medicine. Microtechnology found its first application in electronic, and in a matter of a few decades, revolutionized our daily lives. The concept of miniaturization and functional integration, i.e., the microfabrication of different electronic components and the integration of these components to form complex integrated circuits, is a strategy that can also be used in other fields, e.g., mechanics, optics, chemistry, and life sciences. In 1979, Terry and coworkers presented "a gas chromatography air analyzer fabricated on silicon wafer using integrated circuit technology" [50]. This was the first publication discussing the possible use of techniques borrowed from microelectronics to fabricate a structure for chemical analysis. The introduction of the concept of micro total analysis systems, μTAS, by Manz and coworkers in 1992 triggered rapidly growing interest in the development of microsystems where all the stages of chemical analysis like sample preparation, chemical reactions, analyte separation, analyte purification, analyte detection, and data analysis are performed in an integrated and automated fashion [1]. The realization of such chemical analysis systems requires miniaturization and integration of a wide variety of components, e.g., mechanic, fluidic, optic, and electronic.

Microfabrication, i.e., the fabrication of structures down to micrometers in size, is essential to the development of μTAS. Silicon presented an obvious choice as a material for the microelectronics industry due to its semiconductor properties. The explosive growth of microelectronics has led to a wide range of microfabrication tools for silicon and very high levels of experience and expertise exist for working with this material for microtechnology. Silicon is ideal for microfabrication of electronic, mechanics, and optic components and thereby allows for high levels of functional integration (Microelectromechanical Systems, MEMS). However, the superiority of silicon as a material for μTAS is debatable, because the chemical stability of silicon is not very good. Although the surface of silicon can be treated to withstand harsh chemical environments, other materials may be more suitable for certain biochemical applications. Furthermore the high cost of silicon, especially in single-use applications, where μTAS are in contact with biohazardous materials, requires alternatives. Polymers possess a number of attractive qualities for use in chemical or biochemical microsystems, like optical transparency and chemical resistance to aggressive media. Most importantly, polymers can easily be machined on the micrometer scale using a variety of methods like milling, laser ablation, hot embossing, and injection molding. Furthermore, very persuasive is the very low per-unit manufacturing cost, which is attainable when production is scaled up to batch sizes in the range of hundreds.

Hybrid solutions where microstructures with different functions, fabricated in different materials, are assembled to make up a complete μTAS.

15.5 Acoustically Driven Microfluidics

As pointed out earlier, microfluidic devices for sophisticated lab applications usually comprise different components: First, the small amounts of fluids need to be confined to some kind of containers or reactors, holding specific amounts of the liquid. These containers and reactors are then connected via miniaturized tubes or channels, which are operated by small pumps and valves. The whole system additionally needs to be interfaced to the outside world. The smallness of a microfluidic chip (tube diameters are typically of the order of 100 μm or less), ensures that only tiny amounts of reagents are needed for a chemical or biological reaction, on the other hand, however, it also causes complications that are not of relevance in macroscopic fluid handling systems.

As usual for fluidic problems, one has to first regard the Navier–Stokes-Equation, describing the flow in a hydrodynamic system. It is a nonlinear equation in the velocity components, reading

$$\rho \frac{\partial \vec{v}}{\partial t} + \rho \left(\vec{v} \bullet \mathrm{grad} \right) \vec{v} = -\mathrm{grad}\left(p \right) + \eta \Delta \vec{v} + \vec{f}. \tag{15.1}$$

$\vec{v}$ is the velocity field of the flow, h the viscosity, and r the mass density of the fluid. p denotes the pressure that a fluid element experiences from its surroundings and $\vec{f}$ is an externally applied body force driving the flow.

The term $\rho \left(\vec{v} \bullet \mathrm{grad} \right) \vec{v}$ describes the inertia of the fluid element and $\eta \Delta \vec{v}$ marks the viscous term. The interaction between the fluid confined in a "lab-on-a-chip" and the tube walls leads to hydrodynamic features that are usually given by a single number characterizing the flow behavior in a fluid, the Reynold's number Re, describing the ratio between the inertial and the viscous term:

$$Re = \frac{\rho v^2}{l} \frac{l^2}{\eta v} = \frac{\rho v l}{\eta} \tag{15.2}$$

l denotes a typical length scale in the system under consideration, e.g., the channel diameter. For a microfluidic system, Re is usually a small number, indicating the little importance of inertia in the problem. The most prominent consequence thereof and hence the most important difference to a macroscopic fluid volume is probably the lack of turbulent flows in a microfluidic system. The transition between turbulent and laminar flow usually occurs at a threshold Reynold's number $Re \approx 2{,}000$. Given $l \approx 10$ mm, $v \approx 100$ mm/s, and the material parameters of water ($r = 10^3$ kg/m^3, $h = 10^{-3}$ kg/(msec)), we end up at $Re \approx 0.001$. For a lab-on-a-chip application, this smallness usually causes severe problems. For instance, the mixing of two fluids or stirring a liquid to enhance homogeneity or to speed up a chemical reaction is a very difficult task for a purely laminar flow system. Also, the pumping of a low Reynold's fluid is difficult, as the interaction of the fluid with the vessel walls mimicks a high viscosity. We will show, that many of these obstacles can be overcome by employing our acoustic methods.

15.6 Experimental Details

Flow in a microfluidic system can be achieved by the interaction of the fluid and surface acoustic waves (SAWs), propagating at the surface of a solid chip. The driving force behind this interaction and the resulting acoustically driven flow is an effect called "acoustic streaming". It is a consequence of the pressure dependence of the mass density, ρ, leading to a nonvanishing time average of the acoustically induced pressure. Although acoustic streaming is a well-known effect for a long time in macroscopic, classical systems, little attention has been paid on it, so far, in terms of miniaturization [51].

Here, we report on SAW streaming in a fluid residing directly on top of a planar chip. SAW have been first described in combination with earthquakes [52]. Meanwhile, reduced to the significantly smaller nanoscale, they found their way into much friendlier fields: SAW devices are widely used for RF signal processing and filter applications and have become a huge industry in mobile communication. Also SAW are a vital part in basic research on the nanoscale. Here, the optical and electronic properties of semiconductor nanosysyems have been intensively investigated over the last two decades [53]. SAW are especially convenient to excite on piezoelectric substrates. A well-defined wavelength and frequency can be excited if a specially formed pair of metal electrodes is deposited on top of the substrate. Such electrodes are usually referred to as interdigitated transducers (IDT). A high-frequency signal applied to such an IDT is then converted into a periodic crystal deformation and if fed with the right frequency

$$f = \frac{v_{SAW}}{\lambda} \qquad (15.3)$$

a SAW is launched. Here, v_{SAW} denotes the sound velocity of the respective substrate, and the wavelength, l, is given by the lithografically defined periodicity of the IDT. Typical wavelengths of technically exploited SAW range from about $l = 30$ mm at $f \approx 100$ MHz. If a second IDT was placed downstream the substrate surface, a so-called delay-line would be formed. Both transducers, their design, and the substrate properties thus act as a high-frequency filter with a predetermined frequency response. They are lightweight, relatively simple and of low cost, and can be produced very reproducible, which explains their massive use in high-frequency signal processing like mobile telephony.

Most of the energy propagating in a SAW (usually more than 95%) is of mechanical nature. Viscous materials like liquids absorb a lot of this mechanical energy at the surface of the chip. The interaction between a SAW and a liquid on top of the substrate surface induces internal streaming, and, as we will point out, at large SAW amplitudes, this can even lead to a movement of the liquid as a whole [54, 55]. In Fig. 15.1, we depict the basic interaction between a SAW and a fluid on top of the SAW carrying substrate.

The SAW is approaching from the left and entering the fluid covered region (represented by a droplet in this case) of the chip. There, it becomes attenuated by a

Fig. 15.1 Sketch of the acoustic streaming acting on a small droplet on the surface of a piezoelectric substrate. The acoustic energy is radiated into the fluid under an angle QR, leading to internal streaming in the small fluid volume

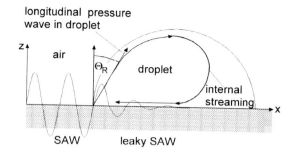

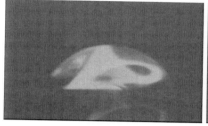

Fig. 15.2 SAW induced internal streaming in a small water droplet (side view, volume is approx. 50 nl). A dried fluorescent dye on the surface of the chip is dissolved by SAW agitation, and rapidly fills the whole droplet volume

viscous damping mechanism, which leads to the excitation of a sound wave in the fluid itself. Phase conservation requires that this sound wave is entering the fluid under an angle Q_R, very much like a diffracted beam in optics. SAW streaming then leads to an internal streaming pattern within the droplet, the exact shape of which is determined by Q_R and the geometry of the fluid volume.

15.7 Acoustic Mixing

The internal streaming as shown in Fig. 15.1 can be very efficiently used for mixing smallest amounts of fluid. In Fig. 15.2, we show a series of two snapshots, about half a second apart, taken for a 50 nl droplet. Here, some fluorescent dye had been deposited at the chip surface just before the water droplet was placed on top. Not only is the dye dissolved by the internal streaming, but is also distributed across the whole volume of the droplet. It should be noted at this point that the SAW-induced streaming still is laminar, as the Reynold's number, Re, is so small. The complex flow pattern, however, strongly supports complex material folding lines, which in turn facilitate a quick mixing [54,55]. Moreover, by switching the SAW frequencies and or directions during the mixing, different material folding lines are generated which further improve mixing.

15.8 Droplet Actuation

For somewhat higher SAW amplitudes, the acoustic streaming effect leads to a strong deformation of the liquid surface and a momentary asymmetry in the wetting angles left and right with respect to the SAW impingement. Especially for small droplets, this leads to a movement and an actuation of the whole droplet into the direction away from the SAW. In this sense, one can regard an IDT on a piezoelectric substrate as an integrated nanopump. Moreover, employing a chemical surface modification, one can define hydrophobic and hydrophillic regions on the chip surface, acting as anchors or fluidic tracks for small droplets. As an example of the versatility of this approach in terms of lab-on-a-chip applications, we show in Fig. 15.3 a series of snapshots of a programmable microfluidic chip with integrated planar pumps.

Here, we have used three different droplets of different chemical solutions, which exhibit a color change when merged and mixed. It should be noted that while the droplets are actuated and eventually merged, in their bulk, the SAW streaming leads to extremely fast chemical reactions compared with a diffusion-only driven process.

15.9 PCR-Chips

DNA amplification (mostly by polymerase chain reaction, PCR [56] and subsequent DNA analysis, including hybridization, sequencing, and genotyping, have been facilitated by the use of microfluidic chips. PCR-microchips were achieved following

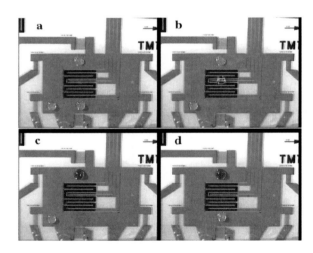

Fig. 15.3 A SAW driven microfluidic processor. Three droplets (approx. 100 nl each) are moved "remotely controlled" and independently by the nanopumps. (**a**) through (**d**) represent a series of subsequently taken snapshots to visualize the movement, and the "nanochemical reactions" occurring when the droplets are merged and mixed by the action of the surface wave. The chip not only contains the nanopumps and the fluidic environment but in the center additional real estate like sensors, and heaters

different approaches. A summary of diverse concepts is reported and reviewed by Kricka and Wilding [57]. Two alternative methods have been reported to achieve temperature variations needed for PCR:

- Stationary On-chip PCR
- Flow-through PCR-chips

Besides well-established materials, like silicon (Si) or glass needed in microsystem technology, also polymers and ceramics are in use, in consideration of the difficulty in forming chambers and channels on glass substrate. Temperature gradients are driven by external resistive heaters or Peltier elements, integrated resistive heaters, or infrared heating. Faster thermal cycling for PCR can be achieved by reducing the thermal mass on a Si-glass chip [58, 59]. For instance, this was accomplished by creating grooves, 1 mm wide, 280 μm deep, in the Si-chip where the thin-film heater was located. As a result, high rates of heating at 36°C/s, and cooling at 22°C/s, were obtained [59]. In one report, a thin-walled PCR chamber was constructed on a PET chip. The thin membrane, 200 μm, between the chamber and the Al heater resulted in only a short 2 s thermal delay. Fast heating, 34–50°C/s, and cooling, 23–31°C/s, rates were obtained [60]. In another report, a thin Si-membrane, ~50 μm, was formed between the Pt thin-film heater and the Si-based PCR chamber. This led to a heating rate of 15°C/s and a cooling rate of 10°C/s [61]. Furthermore, a thin-walled PCR chamber was constructed on a Si-PMMA chip. The Si-membrane between the chamber and heater is $0.8\,\mu m\,SiO_2/0.3\,\mu m\,Si_3N_4/0.8\,\mu m\,SiO_2$. This combination produces a stress-reduced membrane, because SiO_2 is heat-compressive and is Si_3N_4 heat-tensile. This chamber design allows for fast rates of heating at 80°C/s, and cooling at 60°C/s [62]. A noncontact heating method called infrared-mediated temperature control has been employed for PCR. Only the solution absorbs IR, because the chip material, polyimide, does not absorb IR. Therefore, the low thermal mass of the solution allows for fast thermal cycling, and 15 cycles for amplifying a 500 bp long λ phage DNA fragment have been achieved in 240 s [63].

Due to miniaturization, On-chip PCR benefits from the high SVR of the PCR chamber. For instance, SVRs of Si-chambers are $10\,mm^2/\mu l$ [64] and $17.5\,mm^2/\mu l$ [65], which are greater than $1.5\,mm^2/\mu l$ in conventional plastic reaction tubes and $8\,mm^2/\mu l$ in glass capillary reactive tubes [66]. This can lead to the inhibition of PCR reactions on native chip substrates like glass or silicon, due to adsorption of biomolecules to the chip surface [67]. To prevent adsorption of PCR polymerase or nucleic acids, surface treatment and modification or additives in the PCR mixture are necessary.

The basic concepts are described in the subsequent sections.

15.10 Stationary On-Chip PCR

Much effort has been made to integrate PCR chambers on microchips to carry out amplification of DNA molecules prior to their analysis. For instance, PCR was first achieved on a Si-based reaction chamber (25 or 50 μl) integrated with a polysilicon thin-film (2,500 Å thick) heater for the amplification of the 142 bp long GAG gene sequence of HIV, cloned in bacteriophage M13 [68]. In another report, PCR of a 500 bp bacteriophage λDNA fragment was performed in a 10 µl chamber fabricated in a Si-Pyrex chip. Finally, the temperature stayed at 72°C for 5 min. Subsequently, off-chip agarose gel electrophoresis was performed. Nevertheless, it was found that PCR on chip was not efficient as conventional PCR [3]. In one report, anti Taq DNA polymerase antibody was employed to avoid loss of PCR efficiency [64]. The antibody inhibited the Taq polymerase before PCR reagents attained a high temperature, and this procedure is thus called hot-start PCR. In this procedure, the loss of Taq polymerase due to nonspecific binding was reduced. This on-chip hot-start PCR resulted in a more consistent and higher yield than that obtained in the PCR chip without hot-start and even that in the conventional PCR reaction tube (with hot-start) [69].

PCR has been achieved not only from extracted DNA, but also directly from DNA in cells. For instance, the around 100 bp long human CFTR fragment was amplified from isolated human lymphocytes. The results were comparable to PCR from human genomic DNA extracted from the cells. The results indicated that tedious extraction of the DNA template might not be necessary, and PCR can be conducted directly on lysed cells [69]. In a dual-function Si-glass microchip, the isolation of white blood cells (WBC) from whole blood (3.5 µl) using weir-type filters (3.5 µm gap) was followed by PCR of the 202 bp long exon 6 region of the dystrophin gene. Due to the PCR inhibition of hemoglobin, red blood cells are removed and the retained WBC are released by an initial denaturation step at 94°C [64].

Integration of the PCR chamber and subsequent on-chip capillary gel-electrophoresis (CGE) analysis has been reported [70]. First, the target DNA, a 268 bp long β-globin target cloned in M13 or 159 bp long genomic salmonella DNA, was amplified in a microfabricated PCR reactor. The 20 µl reactor was heated by a polysilicon heater (3,000 Å Si doped with boron). Then the PCR products were directly injected into a glass capillary electrophoresis chip for CGE separations. The heating and cooling rates are 10 and 2.5°C/s, respectively, compared with typical rates of 1–3°C/s in conventional thermal cyclers. The use of a thin-film heater permitted a PCR cycle as fast as 30 s in a 280 nl PCR chamber. Stochastic DNA amplification has been demonstrated using one to three 136 bp long templates. Before separation, the PCR mixture was prevented from flowing into the capillary electrophoresis channel using the passive barrier formed by the hydroxyethylcellulose, sieving medium [71]. A Si-based PCR chamber was also interfaced to a glass chip using a PDMS gasket for DNA-based bird sex determination [72].

Another method for fast thermal cycling is noncontact heating via infrared (IR) excitation of the vibrational bands of water. This type of fast thermal cycling was pioneered by the Landers group and demonstrated in capillaries [73] and both polymeric [63] and glass microdevices [74].

However, the concept of microfluidics as a closed system has to deal with some major problems. The pressure required for moving the liquids scales inversely with the channel dimensions [75]. The Wixforth group introduced a microfluidic device

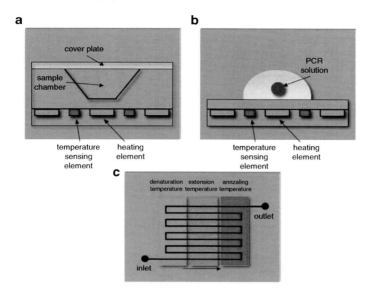

Fig. 15.4 Working principles of PCR chips: (**a**) stationary single chamber on-chip PCR (**b**) virtual reaction chamber (VRC). The PCR sample is introduced into the single or virtual chamber. The chip is then heated and cooled to provide thermocycling conditions; (**c**) a serpentine channel continuous flow-through. The sample is introduced in the inlet and pumped unidirectionally through. Temperature zones are provided by three heaters. Also spiral channel based and bidirectional flow setups are available

operating at a planar surface instead of a closed channel network (VRC, Fig. 15.4b). The fluid is transported in single droplets using surface acoustic waves (SAW) on a piezoelectric substrate [76]. This working principle is the basis of the presented novel lab-on-a-chip system. A thermal convection PCR system, created by a micro immersion heater was introduced by the Braun group [77]. Instead of repetitive heating and cooling, the temperature gradient induces thermal convection, which drives the reaction liquid between hot and cold parts of the chamber. They were able to amplify a 96 bp long λDNA fragment with 500 pg template material.

For complex samples containing several different DNA fragments, multiplex PCR has been carried out. On-chip multiplex PCR was achieved on four DNAs 199–500 bp long fragments of bacteriophage λDNA, a 346 bp *E. coli* genomic DNA fragment, and a 410 bp long *E. coli* plasmid DNA fragment. After PCR, the fluorescent intercalating dye TO-PRO was added to the PCR reservoir, and CGE separation was performed downstream [78].

Unspecific degenerate oligonucleotide-primed PCR (DOP-PCR) with 9.6 ng human genomic DNA as target material was performed before multiplex PCR of the human Dystrophin gene [79].

Besides qualitative also quantitative real-time PCR amplification and detection were also facilitated using microfluidic chips. Real-time monitoring of PCR amplification was achieved by sequential CGE after second intervals [70]. The use of an intercalating dye, SYBR Green I, can only be performed in an all-glass device, rather than one of PDMS-glass, because the dye and DNA appeared to migrate into the PDMS, Sylard 184, polymer [80]. A novel method for DNA amplification and specific sequence detection in an integrated silicon microchamber array was described by the Tamiya group [81]. They were able to amplify and to detect 0.4–12 target copies of Rhesus D gene in each 40 nl microchamber with TaqMan PCR.

Research on the integration of PCR chip designs and subsequent DNA analysis is still a very active field of research. A schematic draw of an On-chip PCR and the operating principle is shown in (Fig. 15.4a, b).

15.11 PCR on a Chip

Having a completely programmable microfluidic chip at hand, one can aim toward more complex assays, e.g., PCR on a chip [82]. Here, however, comparably high temperatures are involved during the protocol. Such high temperatures are not compatible with the "open geometry" of our droplet-based fluidics. To avoid evaporation – especially during the 95°C cycle of a PCR process – the aqueous sample solution is covered with a mineral oil layer.

The sample contains all the "PCR-mix" including template, primers, and polymerase for a successful amplification of a small amount of genetic material. The sample volume in this case is well below $10\,\mu l$, a "virtual liquid test tube" as depicted in Fig. 15.5a). Here, the sample droplet has a red color for better visu-

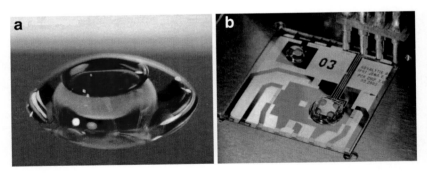

Fig. 15.5 (a) "Virtual liquid test tube" for high temperature application of the planar SAW driven fluidics. To avoid evaporation of the aqueous sample solution (*red*), the droplet has been covered by a thin mineral oil layer. Both represent a fluid test tube for further processing of the sample. (b) Programmable microfluidic biochip for multispot PCR in top view. Apart from SAW driven nanopumps for the fluid actuation, the chip also hosts a heater and a thermometer. The chip is connected to the outside via pogo contacts. Underneath the transparent piezoelectric substrate, a microscope objective is mounted for fluorescence monitoring of the PCR progress employing an intercalating dye

alization. As the programmable biochip is fabricated employing standard planar lithographical technology, we can also include additional functional elements like thermometers and heaters. The optical transparency and the lack of self fluorescence of the substrate materials used add further value to this lab-on-a-chip application [76]. In Fig. 15.5b), we show a typical PCR chip. In the lower right corner, an oil droplet is seen, with in this case four different sample droplets, each holding a different sample. From below, an optical system is attached to monitor the progress of amplification using an intercalating dye. For details of the PCR protocol, the sensitivity of the technology, and typical experimental results we refer the readers to [76].

15.12 Blood Flow on a Chip

The versatility of the acoustically driven planar microfluidic chip can also be nicely demonstrated in a recent study of ours, where we tried to simulate blood flow on a chip. Studying cell or platelet adhesion (hemodynamics) in small capillaries and arteries is a particularly challenging topic, since it is difficult to be mimicked in an in vitro system and hard to access by optical or mechanical means. As the diameter of capillaries and arteries become as small as one micrometer, the Reynold's number becomes very low, making the pumping mechanism extraordinary difficult. This calls for a new pumping principle when designing an in vitro model to study the basic principles taking place during blood flow. Figure 15.6 describes the principle of a microflow chamber, which is directly built on a SAW chip (μ-FCC). This

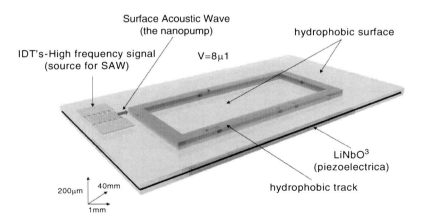

Fig. 15.6 A SAW-driven microfluidic chip. A SAW, excited electrically, eventually interacts with the confined liquid at the solid–liquid interface and drives the liquid to flow (acoustic streaming). The surface wave basically acts as a localized pump, as its mechanical energy is absorbed by the liquid over only a few micrometers. Due to the small scales of this microfluidic system, it creates a homogenous laminar flow (low Reynold's number) along the channel, mimicking the blood flow in arteries or capillaries

system can resemble a type of "artificial blood vessel." The fluid on the chip is confined to a trajectory or a virtual container by a chemical modulation of the surface wettability using hydrophobic/hydrophilic surface functionalization employing soft lithography. The heart of the chip, the nanopump, is driven by a surface acoustic wave (SAW) causing the liquid flow by inducing surface acoustic streaming (see Fig. 15.1). The amplitude of the SAW in the liquid decays exponentially within a characteristic length scale of a few micrometers. Therefore, considering the total length of a typical channel (40 mm), the pump basically acts as a point-like source, driving the liquid to flow according to conservation of mass. The flow chamber has no dead volume, allowing for the investigation of even expensive or rare substrates with a volume 100–1,000 times less than usually required. The particularly small sample volume of only 8 μl is probably the most prominent advantage of our nanopump-driven planar flow chamber chip. Moreover, the chip components are inert and entirely compatible with biological systems and the system is free of all movable parts, making the handling extremely simple. The complete optical transparency of the $LiNbO_3$ substrate allows for individual cells to be tracked over long periods of time, when the μ-FCC is mounted directly onto a fluorescence microscope.

Fluidic lanes ranging from a few μm up to 10 mm can be realized and the open structure on a free surface allows for direct access to the channel at all times (e.g., for the addition of antibodies or drugs). The technique also provides maximum freedom for mimicking all possible vessel architectures and conditions as they may exist in nature with curved, branched, or restricted vessels. Finally, the SAW-based pumping system of our chip guarantees for a homogenous, easily controllable laminar flow the profile of which covers the complete range of physiological flow and shear conditions from 0 to several $10,000\,s^{-1}$ [83]. Finally, no restrictions are put on the surface functionalization. Artificial lipid membranes, protein coats, and even confluent cell layers have been successfully be prepared.

15.13 Proteins Under Flow

Using this set up, we studied the effect of shear flow on a protein called VWF. It is a glycoprotein being synthesized and stored in endothelial cells, and has found to play an important role in blood coagulation, particularly in regions where the shear rates are high. Under normal conditions, it assembles into multimers (biopolymers) that when stretched can become as long as 100 μm. The monomeric length is unusually large (~100 nm) and contains 2,050 amino acids residues. Intuitively, one would expect that blood platelet adhesion is always decreasing with increasing shear force applied to the platelet. Surprisingly, VWF-mediated adhesion, on the other hand, is strongly enhanced under high shear-flow conditions [84]. VWF was spread over the hydrophilic track of the flow chamber chip and exposed to various shear flows. Images of VWF at concentrations typical of human blood ($c \approx 2\,\mu g/ml$) are presented in Fig. 15.6. At shear rates between $\dot{\gamma} \approx 10$ and $1,000\,s^{-1}$, the biopolymer exhibits a compact conformation (Fig. 15.6 left). The size of the VWF globules was estimated within our fluorescence setup to be $d \approx 2\,\mu\text{m}$, clearly showing that it consists of more than one monomer. This compact conformation remains unchanged as long as the shear rate is maintained below a certain (critical) value $\dot{\gamma}_{\text{crit}} \sim 5{,}000\,\text{s}^{-1}$. Increasing the shear rate above $\dot{\gamma}_{\text{crit}}$ induces a shape transformation of the VWF fibers from a collapsed to a stretched conformation of length $l \approx 15\,\mu\text{m}$ (Fig. 15.6 *right*). This transition is reversible, as we observe an immediate relaxation of the protein to its compact conformation when the flow is turned off.

The fact that the transition occurs at such a high shear rate cannot be understood from previous studies on linear chains under good solvent conditions. For example, it has been shown that DNA of roughly the same length as the VWF fibers studied here will exhibit drastic changes of elongation for shear rates as low as $30\,\text{s}^{-1}$ at a viscosity of 1 cP (i.e., water or the phosphate buffer employed here). Our studies, however, indicate that significant changes in conformation occur only at shear rates $\dot{\gamma} \geq 5{,}000\,\text{s}^{-1}$. This value is more than two orders of magnitude higher than that reported for DNA. We assume that strong attractive interactions between monomers hold the VWF fiber tightly together, even under strong shear conditions. Therefore, we propose that a single VWF biopolymer forms a compact or collapsed structure similar to a folded protein (Fig. 15.7 *left*) [7].

This explains the counterintuitive observation mentioned at the beginning of this paragraph. Under high shear flow VWF mediates blood platelet adhesion more effectively. In its collapsed state, all binding sites are buried inside the coil. Once the critical shear is exceeded, VWF stretches into a long thin fiber exposing all its binding sites. If VWF touches the surface in this conformation, which is highly unlikely in small capillaries, it will immediately be immobilized and will serve as a "sticky" grid for free-floating blood platelets. Since forces, conformation, and function are so closely related, we call this "self-organized blood clotting."

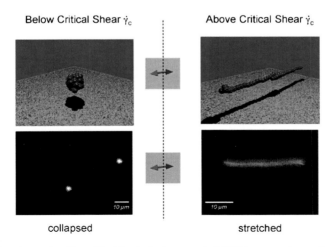

Fig. 15.7 Proteins under Flow. The blood clotting protein VWF is exposed to various shear flow conditions. Only when a critical value is exceeded the protein elongates in a reversible fashion. The unusually high shear rates could not be explained by conventional hydrodynamic models, instead a new theory was developed [85]

15.14 Cell–Cell Interactions on a Chip

Although our planar flow chamber chip represents an elegant technique for examining many relevant topics, it turned out to perform suboptimally for cell culture on its surface: Initial attempts to perform cell culture on a planar flow chamber chip were hindered by the fact that the protein-rich media influenced the channel architecture due to the protein adsorption on the hydrophobic surface. As a result, this led to an unwanted inhomogeneous flow. To avoid this, we developed a more sophisticated 3D version of our system, which easily allows for successful cell culture while retaining all of the advantages of the planar system previously described. We achieved this by a flexible yet stable three dimensional architecture by adding "walls" to the channel employing the fully transparent, biocompatible synthetic elastomer polydimethyl-siloxane (PDMS).

The usefulness and the validity of our system were tested by examining the adhesion of human melanoma cells at various shear rates over time. The results can be seen in Fig. 15.7, which presents our cell culture chip and the PDMS channel. Note that the cells form a confluent layer filling out the entire channel without growing outside of the boundaries. This is evident by the sharp line marking the channel wall. In addition, the cells appear healthy, divide readily, and are securely attached to the chip surface. This cultured layer remains completely unaffected by the SAW, which still induces a continuous flow along the channel.

Finally, we were able to demonstrate the use of our system as a cultured micro flow chamber. The individual cells were tracked at three different points in time. Note that the underlying layer of cells is neither moving nor changing in its overall appearance, clearly indicating its integrity. This experimental finding demonstrates the suitability of our chip to be used as a type of vascular model and further shows the ease with which the flow can be observed, attributable to the transparent design (Fig. 15.8).

15.15 Microdissection

Precise manipulation and microdissection of, e.g., genetic material helps to design chips for genetic analysis, to develop biosensors and lab-on-a-chip diagnostic devices. Nanomanipulation could be defined as the manipulation of nanometer size objects with nanometer size actuator with nanometer precision [86]. Talking about manipulation, it is meant that objects are pushed, pulled, positioned, assembled, cut, etc. by controlling external parameters. Based on the different type of interaction, biological micro/nano-manipulation can be divided into mechanical contact, optical, electrical, fluidic, and electrical noncontact and hybrid systems (see Fig. 15.9). Microdissection of genetic material and isolation of fragments can be accomplished in several ways: (1) using an atomic force microscopy or extended glass needles, (2) by laser microbeams and optical tweezers, (3) an ultrasonically oscillating needle, and (4) by low pressure means. In the following chapter, the different available genetic microdissection techniques are reviewed.

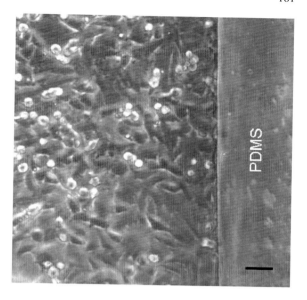

Fig. 15.8 Cell–cell interactions on a Chip. Cells are cultured directly into the microfluidic channel using PDMS walls. The cells form a confluent layer over the entire channel and do not grow across the boundaries (PDMS walls). Cells flowing over the intact and immobilized confluent cell (brighter appearing cells). Thanks to the optical transparency of the piezoelectric substrate and the continuous channel design, the cells can be tracked for an unlimited amount of time. The scale bar represents 50 μm

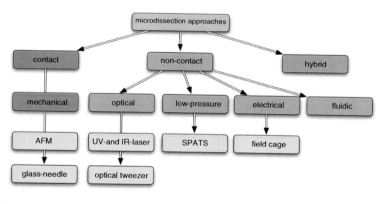

Fig. 15.9 Overview of biological nanomanipulation types

15.16 Extended Glass-Needle Microdissection

Physical dissection of genetic material offers a direct approach to the cloning, amplification, and analysis of DNA sequences that range up to thousands of kilobase pairs in length. Initially microdissection of genetic material was performed with extended glass-needles. This mechanical approach, in which the tip of the needle is in contact with the sample, allows the dissection of material in the range of 1μm. The first experiments, which used extended glass-needles, were performed on polytene chromosomes of Drosophila melanogaster [87]. The methodical steps were transferred to mammalian and human chromosomes [88–90]. The aim of these experiments was the integration of the isolated chromosomal fragments into vectors for subsequent cloning. These experiments were limited by the small amount of DNA available. This limitation was overcome by the development of PCR; [56]. One possibility was the introduction of primer-specific sequences, which were ligated to the microdissected DNA [91, 92]. These highly region-specific probes are extremely valuable for molecular cytogenetic studies as well as for positional cloning projects. With improvements in mechanical microdissection techniques using extended glass needles and PCR, there are two distinct methods for generating a chromosome library from microdissected chromosomal DNA: direct cloning [93] and PCR-mediated cloning [94–96]. Depending on the primer, the latter is divided into degenerate oligonucleotide-primed PCR using random primers [95, 97, 98] and vector- or adaptor-mediated PCR with specific primers [94, 99, 100]. A lot of protocols for the PCR amplification of DNA from few or even single cells have been published over the past years. These include primer-extension preamplification [101], degenerate oligonucleotide-primed PCR [95] and Alu-PCR [102]. They are of variable complexity and none of them has convincingly demonstrated the homogenous amplification of the genome of a diploid cell, although some are quite useful for various cytogenetics analyses [103, 104]. A strategy for rapid construction of whole chromosome painting probes (WCPs) by chromosome microdissection has been developed by Guan et al. [105]. In this approach, WCPs were prepared from 20 copies of each target chromosome microdissected from normal human metaphase chromosomes and then directly amplified by PCR using a universal primer. Different authors reported various numbers of chromosomes needed for the generation of painting probes, ranging from more than 20 [105, 106] to less than 10 [104, 107]. Linker-adaptor PCR can overcome this problem and will be discussed in detail later. Chromosome microdissection and amplification of the isolated fragments by MboI linker-adaptor PCR for genetic disease analysis were described in the early 1990s [108].

The glass needle microdissection is conventionally used for various microoperations like perforation of cells, cutting off and so on. The glass needle is effective for the microoperation, because it is easy to make the tip diameter to be submicro order, but there are defects such as low strength and fragility. Currently, most of such operations are manually done by highly skilled operators under the microscope. The limited visible area at a high magnification of microscope makes it difficult for the operator to trace the target. The operations are tedious and time-consuming.

15.17 Laser-Based Microdissection

The noncontact technique of laser-based microdissection of entire chromosomes and chromosomal fragments was demonstrated [109]. By combining PCR, cloning techniques and laser microdissection, it was possible to generate region specific chromosomal probes for molecular cytogenetics [110–112]. But, the chromosomal fragments had to be collected with an extended glass needle after laser

15 Integrated Lab-on-a-Chip System in Life Sciences

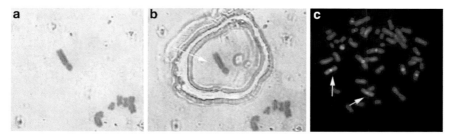

Fig. 15.10 (**a**) and (**b**) laser based microdissection and isolation of a single metaphase chromosome fixed on a supporting membrane, *see arrow*; (**c**) human chromosome 3 specific paint probe generated from a single isolated chromosome, *see arrows*

microdissection. Newer developments include laser micromanipulation and dissection, which has considerably increased the ease and speed to isolate chromosomes [113]. The generation of chromosomal-specific painting probes was reported with an entire noncontact laser-based microdissection in single and multicolor-FISH experiments [114–116] (see Fig. 15.10). By combining linker-adapter PCR and comparative genome hybridization, it was also possible to detect loss of heterozygosity in single isolated cells [117]. Linker-adaptor PCR utilizes specific linkers ligated to the ends of DNA fragments, generated by restriction enzyme digestion; subsequently, DNA is amplified using PCR primers homologous to the linker-adapter oligonucleotide.

15.18 Atomic Force Microscopy Microdissection

Since the invention of the AFM [118] and its use in structural biology of chromosomes, research has been focused on the use of the AFM not only as an imaging but also as a manipulation tool. By combining high structural resolution with the ability to control the image parameters at any position within the scan area, it is possible to use the AFM as a micromanipulation tool. Hoh et al. demonstrated the possibility of using the AFM as a microdissection device [119]. They performed microdissection on gap junctions between cells. Controlled nanomanipulation of biomolecules was performed on genetic material [120]. Fragments of about 100–150 nm were cut out of circular plasmid DNA. Isolated DNA adsorbed on a mica surface was dissected in air [120–122] and in liquids, e.g., propanol [120], by increasing the applied force to about 5 nN at the AFM specimen. These experiments demonstrated the feasibility of microdissection in the nanometer range. Combining AFM imaging and microdissection, the organization of bovine sperm nuclei was observed and showed small protein and DNA containing subunits with diameter of 50 to 100 nm [123]. A tobacco mosaic virus was dissected and displaced on a graphite surface to record the mechanical properties of the virus binding [124].

Chromosomal dissection allows direct isolation from selected regions. Therefore, it can be used to build chromosome band libraries [91] for cytogenetic mapping strategies or specific cloning projects. AFM microdissection of genetic material in different condensation status, like polytene chromosomes of Drosophila melanogaster, was performed by the group of Henderson [125, 126]. The achieved cut size in chromosomal regions was 107 nm. Depending on the shape of the AFM tip, the size increased to 170 nm in larger regions. Also human metaphase chromosomes were microdissected and the extracted material was used for subsequent biochemical reactions [103, 127, 128]. Manipulation of mouse chromosomes was described. After dissection with a modified AFM tip the collected material was amplified with subsequent southern hybridization of the extracted single-copy DNA [127]. AFM microdissection in a dynamic mode for the chemical and biological analysis of tiny chromosomal fragments was shown. In this approach, the marker gene of the nucleolar organizing region (NOR) was amplified by designed primers for the 5.8S ribosomal DNA after performing a series of single-line scan microdissections. The dissected chromosomal fragments were collected in a second step with a conventional microcapillary [128]. Human metaphase chromosomes were dissected at selected regions by upstream noncontact imaging of the GTG-banded metaphase chromosomes. The microdissection process can be documented [103, 129]. In this direct approach, the extracted genetic material, adhering to the tip, is amplified by unspecific PCR. Then it can be used as a probe for fluorescence in situ hybridization (FISH) [103]. As described earlier, AFM can also be operated in liquid environment. However, microdissection in liquids produces only uncontrolled cuts on rehydrated chromosomes [130].

The combination of high-resolution imaging and manipulation allows for the first time identification of the sample area, microdissection, and nanoextraction of genetic material at once (Fig. 15.11). This nanometer-sized material can be used for further biomedical and biochemical studies (for review [131]).

15.19 Acoustically Driven Cytogenetic Lab-on-a-Chip

While DNA-chips become commercially important, scientific and technical development in the last years generated different approaches of multiparameter tests particular for medical applications, so-called "lab-on-a-chip" (LOC) systems. Miniaturization of analysis systems will yield in an enormous cost-saving in regard to materials like test tubes or microtitre plates as well as biochemical reagents. Furthermore, a smaller sample volume implies in the end a higher sensitivity and homogeneity of detection. In addition, in comparison with serial single analysis, parallelization of analysis enables an enormous time saving due to automation. These micro- and nanolaboratories in the size of a computer chip are equipped with all components necessary for cytogenetic analysis; they are portable, easy to use, flexible, inexpensive, biocompatible, and like computer chips, full programmable.

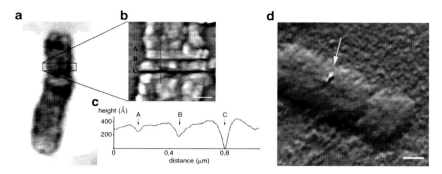

Fig. 15.11 (a) GTG banded human chromosome 9 imaged in noncontact mode, the frame marks the area of a series of microdissections; (b) The chromosomal part was imaged by AFM in ambient conditions after a series of dissections made by AFM. For dissection, z-modulation (∼5 nm) has been used. The oscillation amplitude of the cantilever was smaller than 1% of the amplitude of free oscillation for all cuts. Each cut was performed by scanning one line scan at 1 μm/s with a setup loading force: #A: 7 μN, #B: 9 μN, #C: 13 μN. bar: 500 nm (c) cross sectional analysis along the *red line* indicated in b; (d) AFM microdissection of a single chromatid arm, *see arrow*. The achieved cut size is around 60 nm; bar 1 μm

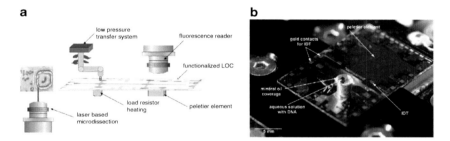

Fig. 15.12 (a) The modular lab-on-a-chip system consists of several units for isolating, processing and analyzing of minute amounts of sample material: laser-based microdissection is followed by processing of the extracted material and detection of hybridized probes or amplified material by a fluorescence reader. All operations on the LOC are controlled by SAW actuated microfluidics. (b) SAW driven lab-on-a-chip system with ten interdigital transducers and two heaters. Transport of minute amounts of sample material in "virtual beakers" is actuated by surface acoustic waves generated via interdigital transducers. The liquid phase comprising the genetic material (*red*) is covered by a thin layer of mineral oil avoiding evaporation. A load resistor heating and a peltier element provide for precise temperature profiles required for PCR methods

Here we present an acoustic driven lab-on-a-chip for cytogenetic and forensic applications (see Fig. 15.12; [132]). In contrast to many other lab-on-a-chip approaches, the fluidic handling is done on the planar surface of this chip, the fluids being confined in "virtual" reaction chambers and "virtual" test tubes in the form of free droplets. The droplets, fluidic tracks and reaction sites are defined at the chip surface by a monolayer chemical modification of the chip surface. In comparison with conventional closed microfluidic systems with external pumping, afflicted with the difficulty to further miniaturize, SAWs are employed to agitate and actuate these little virtual test tubes along predetermined trajectories. These SAWs propagate on a substrate surface, to move and mix smallest fluidic volumina. Liquid amounts in the range from 1 micro- down to 100 picolitre are precisely moved on monolayers of thin, chemical processed fluidic "tracks" without any tubing system. The SAWs are generated by high-frequency electrical impulses on microstructured interdigital transducers embedded into the lab-on-a-chip.

Well-defined analyses, controlled in the submicrolitre regime, can be quickly and gently conducted on the lab-on-a-chip. Apart from its nearly unlimited applicability for many different biological assays, its programmability and extremely low manufacturing costs are another definite advantage of this system. In fact, they can be made so cheap that their use as disposables in many areas of diagnosis can be envisioned. Minute amount of sample material is extracted by laser-based microdissection out of, e.g., histological sections [113, 116]. A few picogram of genetic material are isolated and transferred via a low-pressure transfer system onto the lab-on-a-chip [133]. Subsequently, the genetic material inside single droplets, which behave like "virtual" beakers, is transported to the reaction and analysis centres on the chip surface via SAWs, probably best known from their use as high-frequency filters in mobile phones. At these "biological reactors," the genetic material is processed, e.g., amplified via PCR methods, and genetically characterized [134].

15.20 Summary

We described an unconventional technique to manipulate smallest amounts of liquid on a chip. Employing SAWs on a piezoelectric substrate, we are able to actuate individual droplets along predetermined trajectories, or induce acoustically driven internal streaming in the fluid. This internal acoustic streaming can efficiently be used to agitate, mix, and stir very small liquid volumes, where the low Reynold's number usually only allows for diffusive mixing. As for typical applications, we described a programmable microfluidic chip for droplet-based assays, to perform high-resolution microliter PCR. The technique is equally well suited to actuate or agitate small amounts of liquids either in closed volumes or in an open, droplet based geometry. An example of such a closed though "open" volume is given in the context of a model for the investigation of blood flow on a chip. Here, we showed the feasibility to even grow a cell culture on predetermined areas of the chip, representing a flat version of a blood vessel. The shear force-induced mechanical activation of an important biopolymer for wound healing and cell adhesion is one of our latest unpreceded applications of our quite universal SAW fluidic chip. The combination of the SAW actuated droplet-based fluid handling and SAW-driven fluidics in closed volumes opens a wide field of many different applications.

Acknowledgments This work would have been impossible without the combined efforts of many people involved, both at the Helmholtz Zentrum Munich, the University of Augsburg, and at Advalytix AG. Some of them are listed in the references and others need to be left unaccounted for in this article because of limited space. Their valuable contributions, however, are gratefully acknowledged, in any case. The financial support was provided by the German minister for research and technology (BMBF), the Bavarian Science Foundation (BayFo), the German Science foundation DFG under SPP 1164, and SFB 486, and in part by the German cluster of Excellence "Nanosystems Initiative Munich NIM". M.S. thanks the Fonds der Chemischen Industrie for financial support.

References

1. A. Manz, N. Graber, H.M. Widmer, Sens. Actuat. B Chem. **1**, 244 (1990)
2. A. Manz, Y. Miyahara, J. Miura, Y. Watanabe, H. Miyagi, K. Sato, Sens. Actuat. B Chem. **1**, 249 (1990b)
3. P. Wilding, M.A. Shoffner, L.J. Kricka, Clin. Chem. **40**, 1815 (1994)
4. D.J. Harrison, K. Fluri, K. Seiler, Z. Fan, C.S. Effenhauser, A. Manz, Science **261**, 895 (1993)
5. L. Martynova, L.E. Locascio, M. Gaitan, G.W. Kramer, R.G. Christensen, A. William, W.A. MacCrehan, Anal. Chem. **69**, 4783 (1997)
6. D.C. Duffy, J.C. McDonald, O.J.A. Schueller, G.M. Whitesides, Anal. Chem. **70**, 4974 (1998)
7. S.W. Schneider, S. Nuschele, A. Wixforth, C. Gorzelanny, A. Alexander-Katz, R.R. Netz, M.F. Schneider, PNAS **104**, 7899 (2007)
8. H. Matsuoka, T. Komazakia, Y. Mukaia, M. Shibusawaa, H. Akanea, A. Chakic, N. Uetakec, M. Saito, J. Biotech. **116**, 185 (2005)
9. A. Ashkin, J.M. Dziedzic, Science **235**, 1517 (1987)
10. T.N. Buican, M.J. Smyth, H.A. Crissman, G.C. Salzman, C.C. Stewart, J.C. Martin, Appl. Opt. **26**, 5311 (1987)
11. T. Schnelle, R. Hagedorn, G. Fuhr, S. Fiedler, T. Müller, Biochim. Biophys. Acta **1157**, 127 (1993)
12. C. Duschl, P. Geggier, M. Jäger, M. Stelzle, T. Müller, T. Schnelle, G.R. Fuhr, in *Micro and Nanotechnologies for Life Scienc*, ed. by H. Andersson, A. van den Berg (Kluwer, Dordrecht, The Netherlands, 2005)
13. T. Thorsen, S.J. Maerkl, S.R. Quake, Science **298**, 580 (2002)
14. A. Khademhosseini, J. Yeh, S.Y. Jon, G. Eng, K.Y. Suh, J. Burdick, R. Langer, Lab Chip **4**, 425 (2004)
15. H. Andersson, A. van den Berg, Lab Chip **4**(2), 98 (2004)
16. W.C. Chang, L.P. Lee, D. Liepmann, Lab Chip **5**(1), 64 (2005)
17. N. Xu, Y. Lin, S.A. Hofstadler, D. Matson, C.J. Call, R.D. Smith, Anal. Chem. **70**(17), 3553 (1998)
18. J. Bergkvist, S. Ekström, L. Wallman, M. Löfgren, G. Marko-Varga, J. Nilsson, T. Laurell, Proteomics **2**(4), 422 (2002)
19. S. Fiedler, S.G. Shirley, T. Schnelle, G. Fuhr, Anal. Chem. **70**(9), 1909 (1998)
20. A.Y. Fu, H.P. Chou, C. Spence, F.H. Arnold, S.R. Quake, Anal. Chem. **74**(11), 2451 (2002)
21. A. Wolff, I.R. Perch-Nielsen, U.D. Larsen, P. Friis, G. Goranovic, C.R. Poulsen, J.P. Kutter, P. Telleman, Lab Chip **3**(1), 22 (2003)
22. M.M. Wang, E. Tu, D.E. Raymond, J.M. Yang, H. Zhang, N. Hagen, B. Dees, E.M. Mercer, A.H. Forster, I. Kariv, P.J. Marchand, W.F. Butler, Nat. Biotechnol. **23**(1), 83 (2004)
23. M. Ozkan, M. Wang, C. Ozkan, R.A. Flynn, S. Esener, J. Biomed. Microdev. **5**, 47 (2003)
24. D.R. Reyes, D. Iossifidis, P. Auroux, A. Manz, Anal. Chem. **74**(12), 2623 (2002)
25. P.A. Auroux, D. Iossifidis, D.R. Reyes, A.M. Manz, Anal. Chem. **78**, 3887 (2007)
26. P.S. Dittrich, K. Tachikawa, A. Manz, Anal. Chem. **78**, 3887 (2007)
27. N.T. Nguyen, X. Huang, T.K. Chuan, Trans. ASME **124**, 384 (2002)
28. P.M. White, *Fluid Mechanics* (Mc Graw Hill, New York, 1986)
29. C.H. Ahn, M.G. Allen, in *IEEE 8th Int. Workshop on MEMS (MEMS'95)*, 1995, p. 408
30. J. Doepper, M. Clemens, W. Ehrfeld, S. Jung, P.S. Dittrich, K. Tachikawa, A. Manz, Anal. Chem. **78**, 3887 (2007)

31. S. Haeberle, T. Brenner, R. Zengerle, J. Maeda, Sens. Actuat. A **93**, 266 (2001)
32. Z. Yang, S. Matsumotot, H. Goto, M. Matsumoto, R. Maeda, Sens. Actuat. A **93**, 266 (2001)
33. R.M. Moroney, R.M. White, R.T. Howe, in *IEEE 4th Int. workshop on MEMS (MEMS '91)*, 1991, p. 277
34. N.T. Nguyen, A.H. Meng, J. Black, R.M. White, Sens. Actuat. A **79**, 115 (2000)
35. M. Kurosawa, T. Watanabe, T. Higuchi, in *IEEE 8th Int. Workshop on MEMS (MEMS'95)*, 1995, p. 25
36. A. Wixforth, Superlatt. Microstruct. **33**, 389 (2003)
37. S. Shoji, M. Esashi, J. Micromech. Microeng. **4**, 157 (1994)
38. S. Haeberle, R. Zengerle, Lab Chip **7**, 1094 (2007)
39. S.F. Bart, L.S. Tavrow, M. Mehregany, J.H. Lang, Sens. Actuat. A **21–23**, 193 (1990)
40. G. Fuhr, T. Schnelle, B. Wagner, J. Micromech. Microeng. **4**, 217 (1994)
41. J.R. Webster, M.A. Burns, D.T. Burke, C.H. Mastrangelo, in *IEEE 13th Int. Workshop on MEMS (MEMS'00)*, 2000, p. 306
42. H. Tagagi, R. Maeda, K. Ozaki, M. Parameswaran, M. Metha, Proc. Micro Mechatronics Hum. Sci. **94**, 199 (1994)
43. T.K. Jun, C.J. Kim, J. Appl. Phys. **83**, 5658 (1998)
44. J. Lee, C.J. Kim, in *IEEE 11th Int. Workshop on MEMS (MEMS'98)*, 1998, p. 538
45. S. Boehm, W. Olthuis, P. Bergveld, IEEE 13th Int. Workshop on P.A. Auroux, D. Iossifidis, D.R. Reyes, A. Manz, Anal. Chem. **74**(12), 2637 (2002)
46. A.V. Lemoff, A.P. Lee, Sens. Actuat. **63**, 178 (2000)
47. J.E. Leland, US patent 5005639 (1991)
48. C. Yamahata, M. Chastellain, V.K. Parashar, A. Petri, H. Hofmann, M.A.M. Gijs, J. Microelectromech. Syst. **14**, 96 (2005)
49. T. Ohashi, H. Kuyama, N. Hanafusa, Y. Togawa, Biomed. Microdev. **9**, 695 (2007)
50. S.C. Terry, J.H. Jerman, J.B. Angell, IEEE Trans. Electron Dev. **26**(12), 1880 (1979)
51. C. Eckart, Phys. Rev. **73**, 68 (1948)
52. L. Rayleigh, Philos. Mag. **10**, 364 (1905)
53. A. Wixforth, Int. J. High Speed Electr. Syst. **10**(4), 1193 (2000)
54. A. Wixforth, C. Strobl, C. Gauer, A. Toegl, J. Scriba, Z. von Guttenberg, Anal. Bioanal. Chem. **379**, 982 (2004)
55. T. Frommelt, M. Kostur, M. Wenzel-Schäfer, P. Talkner, P. Hänggi, A. Wixforth, Phys. Rev. Lett. **100**(3), 034502 Epub (2008)
56. K. Mullis, F. Faloona, S. Scharf, R. Saiki, G. Horn, H. Erlich, Quant. Biol. **51**, 263 (1986)
57. L.J. Kricka, P. Wilding, Anal. Bioanal. Chem. **377**(5), 820 (2003)
58. J.H. Daniel, D.F. Moore, S. Iqbal, R.B. Millington, C.R. Lowe, D.L. Leslie, M.A. Lee, M.J. Pearce, Sens. Actuat. A **71**, 81 (1998)
59. D.S. Yoon, Y.S. Lee, Y. Lee, H.J. Cho, S.W. Sung, K.W. Oh, J.H. Cha, G. Lim, J. Micromech. Microeng. **12**, 813 (2002)
60. Q.B. Zou, Y.B. Miao, Y. Chen, U. Sridhar, C.S. Chong, T.C. Chai, Y. Tie, C.H.L. The, T.M. Lim, C.K. Heng, Sens. Actuat. A **102**, 114 (2002)
61. Z. Zhao, Z. Cui, D.F. Cui, S.H. Xia, Sens. Actuat. A **108**, 162 (2003)
62. D.S. Lee, S.H. Park, H. Yang, K.H. Chung, T.H. Yoon, S.J. Kim, K. Kim, Y.T. Kim, Lab Chip **4**(4), 401 (2004)
63. B.C. Giordano, J. Ferrance, S. Swedberg, A.F.R. Huhmer, J.P. Landers, Anal. Chem. **291**(1), 124 (2001)
64. P. Wilding, L.J. Kricka, J. Cheng, G. Hvichia, Anal. Bio-Chem. **257**, 95 (1998)
65. J. Cheng, M.A. Shoffner, G.E. Hvichia, L.J. Kricka, P. Wilding, Nucleic Acids Res. **24**, 380 (1996)
66. M.A. Shoffner, J. Cheng, G.E. Hvichia, L.J. Kricka, P. Wilding, Nucleic Acids Res. **24**, 375 (1996)
67. S.H. Kang, M.R. Shorteed, E.S. Yeung, Anal. Chem. **73**(6), 1091 (2001)
68. M.A. Northrup, M.T. Ching, R.M. White, R.T. Watson, Transducer 924 (1993)
69. J. Cheng, M.A. Shoffner, G.E. Hvichia, L.J. Kricka, P. Wilding, Nucleic Acids Res. **24**, 380 (1996)

70. A.T. Woolley, D. Hadley, P. Landre, A.J. de Mello, R.A. Mathies, M.A. Northrup, Anal. Chem. **68**, 4081 (1996)
71. E.T. Lagally, C.A. Emrich, R.A. Mathies, Labchip **1**, 102 (2001)
72. I. Rodriguez, M. Lesaicherrre, Y. Tie, Q.B. Zou, C. Yu, J. Singh, L.T. Meng, S. Uppili, S.F.Y. Li, P. Gopalakrishnakone, Z.E. Selvanayagam, Electrophoresis **24**, 172 (2003)
73. R.P. Oda, M.A. Strausbauch, A.F. Huhmer, N. Borson, S.R. Jurrens, J. Craighead, P.J. Wettstein, B. Eckloff, B. Kline, J.P. Landers, Anal. Chem. **70**(20), 4361 (1998)
74. C.J. Easley, J.M. Karlinsey, J.P. Landers, Lab Chip **6**, 601 (2006)
75. J.P. Brody, P. Yager, R.E. Goldstein, R.H. Austin, Biophys. J. **71**(6), 3430 (1996)
76. Z. Guttenberg, H. Muller, H. Habermuller, A. Geisbauer, J. Pipper, J. Felbel, M. Kielpinski, J. Scriba, A. Wixforth, Lab Chip **5**(3), 308 (2005)
77. M. Hennig, D. Braun, APL **87**, 183901 (2005)
78. L.C. Waters, S.C. Jacobson, N. Kroutchinia, J. Khandurina, R.S. Foote, J.M. Ramsey, Anal. Chem. **70**, 5172 (1998)
79. J. Cheng, L.C. Waters, P. Fortina, G. Hvichia, S.C. Jacobson, J.M. Ramsey, L.J. Kricka, P. Wilding, Anal. Chem. **257**(2), 101 (1998)
80. H. Mao, M.A. Holden, M. You, P.S. Cremer, Anal. Chem. **74**(19), 5071 (2002)
81. Y. Matsubara, K. Kerman, M. Kobayashi, S. Yamamura, Y. Morita, E. Tamiya, Bioelectronics **20**, 1482 (2005)
82. A. Toegl, R. Kirchner, C. Gauer, A. Wixforth, J. Biomol. Tech. **14**(3), 197 (2003)
83. G.A. Trusky, F. Yuan, D.F. Katz, *Transport Phemomena in Biological Systems Upper Saddle River* (Pearson Education, New Jersey, 2004)
84. Z.M. Ruggeri, J. Thromb. Haemost. **1**(7), 1335 (2003)
85. A. Alexander-Katz, M.F. Schneider, S.W. Schneider, A. Wixforth, R.R. Netz, Phys. Rev. Lett. **97**, 138101. Epub (2006)
86. M. Sitti, in *Proc. of the IEEE-Nanotechnology Conference*, 2001, p. 75
87. F. Scalenghe, E. Turco, J.E. Edström, V. Pirrotta, M. Melli, Chromosoma **82**, 205 (1981)
88. D. Röhme, H. Fox, B. Hermann, A.M. Frischauf, J.E. Edström, P. Mains, L.M. Silver, H. Lehrach, Cell **36**, 783 (1984)
89. A.J. Greenfield, S.D.M. Brown, Genomics **1**, 153 (1987)
90. G. Senger, H.J. Lüdecke, B. Horsthemke, U. Claussen, Hum. Genet **84**, 507 (1990)
91. H.J. Lüdecke, G. Senger, U. Claussen, B. Horsthemke, Nature **338**, 348 (1989)
92. D.C. Johnson, Genomics **6**, 243 (1990)
93. E.M.C. Fisher, J.S. Cavanna, S.D.M. Brown, Proc. Natl. Acad. Sci. USA **82**, 5846 (1985)
94. C. Jung, U. Claussen, B. Horstemke, F. Fischer, R.G. Herrmann, Plant Mol. Biol. **20**, 503 (1992)
95. H. Telenius, A.H. Pelmear, A. Tunnacliffe, N.P. Carter, A. Behmel, M.A. Ferguson-Smith, M. Nordednskjold, R. Pfragner, B.A. Ponder, Genes Chromos. Cancer **4**(3), 257 (1992)
96. N. Stein, N. Ponelies, T. Musket, M. McMullen, G. Weber, Plant J. **13**, 281 (1998)
97. U. Pich, A. Houben, J. Fuchs, A. Meister, I. Schubert, Mol. Gen. Genet. **243**, 173 (1994)
98. B. Liu, G. Segal, J.M. Vega, M. Feldman, S. Abbo, Plant J. **11**, 959 (1997)
99. Q. Chen, K. Armstrong, Genome **38**, 706 (1995)
100. Y. Zhou, Z. Hu, B. Dang, H. Wang, X. Deng, L. Wang, Z. Chen, Chromosoma **108**, 250 (1999)
101. L. Zhang, X. Cui, K. Schmitt, R. Hubert, W. Navidi, N. Arnheim, Proc. Natl. Acad. Sci. USA **89**, 5847 (1992)
102. D.L. Nelson, S.A. Ledbetter, L. Corbo, M.F. Victoria, R. Ramirez Solis, T.D. Webster, D.H. Ledbetter, C.T. Caskey, Proc. Natl. Acad. Sci. USA **86**, 6686 (1989)
103. S. Thalhammer, R. Stark, S. Müller, J. Wienberg, W.M. Heckl, J. Struct. Biol. **119**(2), 232 (1997)
104. J. Weimer, M.R. Koehler, U. Wiedemann, P. Attermeyer, A. Jacobsen, D. Karow, M. Kiechle, W. Jonat, N. Arnold, Chromos. Res. **9**, 395 (2001)
105. X.Y. Guan, P.S. Meltzer, J.M. Trent, Genomics **22**(1), 101 (1994)
106. H.X. Deng, K. Yoshiura, R.W. Dirks, N. Harada, T. Hirota, K. Tsukamoto, Y. Jinno, N. Niikawa, Hum. Genet. **89**(1), 13 (1992)

107. J. Müller-Navia, A. Nebel, E. Schleiermacher, Hum. Genet. **96**(6), 661 (1995)
108. F.T. Kao, J.W. Yu, Proc. Natl. Acad. Sci. USA **88**, 1844 (1991)
109. S. Monajembashi, C. Cremer, T. Cremer, J. Wolfrum, K.O. Greulich, Exp. Cell Res. **167**, 262 (1986)
110. N. Ponelies, E.K.F. Bautz, S. Monajembashi, J. Wolfrum, K.O. Greulich, Chromosoma **98**, 351 (1989)
111. C. Lengauer, A. Eckelt, A. Weith, N. Endlich, N. Ponelies, P. Lichter, K.O. Greulich, T. Cremer, Cytogenet. Cell Genet. **56**, 27 (1991)
112. W. He, Y. Liu, M. Smith, M.W. Berns, Microsc. Microanal. **3**, 47 (1997)
113. S. Thalhammer, G. Lahr, A. Clement-Sengewald, W.M. Heckl, R. Burgemeister, K. Schütze, J. Laser Phys. **13**(5), 681–692 (2003)
114. L. Schermelleh, S. Thalhammer, T. Cremer, H. Pösl, W.M. Heckl, K. Schütze, M. Cremer, BioTech. Int. **27**, 362 (1999)
115. S. Kubickova, H. Cernohorska, P. Musilova, J. Rubes, Chromos. Res. **10**, 571 (2002)
116. S. Thalhammer, S. Langer, M.R. Speicher, W.M. Heckl, J.B. Geigl, Chromos. Res. **12**, 337 (2004)
117. C.A. Klein, O. Schmidt-Kittler, J.A. Schardt, K. Pantel, M.R. Speicher, G. Riethmüller, Proc. Natl. Acad. Sci. USA **96**, 4494 (1999)
118. G. Binnig, C.F. Quate, C. Gerber, Phys. Rev. Lett. **56**, 930 (1986)
119. J.H. Hoh, R. Lal, S.A. John, J.P. Revel, M.F. Arnsdorf, Science **253**, 1405 (1991)
120. H.G. Hansma, J. Vesenka, C. Siegerist, G. Kelderman, H. Morrett, R.L. Sinsheimer, V. Elings, C. Bustamante, P.K. Hansma, Science **256**, 1180 (1992)
121. J. Vesenka, M. Guthold, C.L. Tang, D. Keller, E. Delaine, C. Bustammante, Ultramicroscopy **42–44**, 1243 (1992)
122. B. Geissler, F. Noll, N. Hampp, Scanning **22**, 7 (2000)
123. M.J. Allen, C. Lee, J.D. Lee IV, G.C. Pogany, M. Balooch, W.J. Siekhaus, R. Balhorn, Chromosoma **102**, 623 (1993)
124. M.R. Falvo, S. Washburn, R. Superfine, M. Finch, F.P. Brooks, V. Chi, R.M. Taylor, Biophys. J. **72**, 1396 (1997)
125. C. Mosher, D. Jondle, L. Ambrosio, J. Vesenka, E. Henderson, Scanning Microsc. **8**(3), 491 (1994)
126. D.M. Jondle, L. Ambrosio, J. Vesenka, E. Henderson, Chromos. Res. **3**, 239 (1995)
127. X.M. Xu, A. Ikai, Biochem. Biophys. Res. Commun. **248**, 744 (1998)
128. S. Iwabuchi, T. Mori, K. Ogawa, K. Sato, M. Saito, Y. Morita, T. Ushiki, E. Tamiya, Arch. Histol. Cytol. **65**(5), 473 (2002)
129. S. Thalhammer, U. Köhler, R. Stark, W.M. Heckl, J. Microsc. **202**(3), 464 (2001)
130. R. Stark, S. Thalhammer, J. Wienberg, W.M. Heckl, Appl. Phys. **A66**, 579 (1998)
131. S. Thalhammer, W.M. Heckl, Cancer Genomics Proteomics **1**, 59 (2004)
132. S. Thalhammer, Z. von Guttenberg, U. Koehler, A. Zink, W.M. Heckl, T. Franke, H.G. Paretzke, A. Wixforth, GenomXPress **1**, 07, 29 (2007)
133. D. Woide, V. Mayer, T. Wachtmeister, N. Höhn, A. Zink, U. Köhler, S. Thalhammer. Single particle adsorbing transfer system. Biomedical Microdevices **11**(3), 609–614, (2009)
134. D. Woide, V. Mayer, T. Neumaier, T. Wachtmeister, H.G. Paretzke, Z. von Guttenberg, A. Wixforth, S. Thalhammer, in *Proceedings of the First International Conference on Biomedical Electronics and Devices*, Vol. 2, 2008, p. 265

Part V
Philosophical Aspects of Nanoscience

Chapter 16
Methodological Problems of Nanotechnoscience[1]

V.G. Gorokhov

Abstract Recently, we have reported on the definitions of nanotechnology as a new type of NanoTechnoScience and on the nanotheory as a cluster of the different natural and engineering theories. Nanotechnology is not only a new type of scientific-engineering discipline, but it evolves also in a "nonclassical" way. Nanoontology or nano scientific world view has a function of the methodological orientation for the choice the theoretical means and methods toward a solution to the scientific and engineering problems. This allows to change from one explanation and scientific world view to another without any problems. Thus, nanotechnology is both a field of scientific knowledge and a sphere of engineering activity, in other words, NanoTechnoScience is similar to Systems Engineering as the analysis and design of large-scale, complex, man/machine systems but micro- and nanosystems. Nano systems engineering as well as Macro systems engineering includes not only systems design but also complex research. Design orientation has influence on the change of the priorities in the complex research and of the relation to the knowledge, not only to "the knowledge about something", but also to the knowledge as the means of activity: from the beginning *control and restructuring of matter* at the nano-scale is a necessary element of nanoscience.

Introduction

The sphere of scientific-technological disciplines, which are intensively studied today, along with the natural-scientific, mathematical, social disciplines and humanities, encompasses a great number of varied fields of research, engineering, and design. At present, scientists are founding organizations (a specific range of publications and a limited research community) for scientific research. In addition, as shown earlier, by the second half of the twentieth century, a majority of the

---

[1] This article has been prepared in the scope of RFFI-Project "Technoscience in Knowledge Society: Methodological Problems of the Development of the Theoretical Investigations in Engineering Sciences" No 09-06-00042a.

scientific-technological disciplines began their own theoretical studies, which have received the status of a technical theory. Today, we have, in the scientific community, more connection between science and technology (also in the basic research sphere). We are already talking about "technoscience". In the modern scientific landscape, we can see a yet-more special type of scientific discipline – a scientific-technological discipline. New scientific-technological disciplines are unique in that they emerge at the interface between the scientific and engineering activities and are supposed to ensure an effective interaction of the two aforementioned types of activity. Characteristic of the scientific-technological disciplines is a closer relationship with the engineering practice.

16.1 Different Definitions of Nanotechnology

Nanotechnology is:

1. *a sphere of the scientific and engineering activity* that is connected with

   - organization of the process of creation, fabrication, implementation, use, and development of the nano-scale systems, that is, coordination between the various design tasks and cooperation of the different specialists who solved these tasks
   - support to assemble and to integrate the heterogeneous parts of the designed nanosystem into the organic whole

2. *a sphere of knowledge*, a complex scientific and engineering discipline that integrates

   - means, methods, operations, and procedures of design and research of the nano-scale systems
   - methods and principles of the organization of the scientific and engineering activity
   - knowledge and methods of the modern mathematical, technological, natural, and another sciences that are used for analysis and design of the nano-scale systems and for the organization of the scientific and engineering activity

3. *a concrete-methodological position* that is connected with holistic investigation of the nano-scale systems and of the process of their research, generation, implementation, and fabrication from the cybernetics (algorithmic) and systems approach point of view

One of these definitions, the so-called 'real' definition, which is relevant to the already existing and proposed applications in different fields, refers to a list of particular cases of current research topics. "Such lists typically include scanning probe microscopy, nanoparticle research, nanostructured materials, polymers and composites, ultrathin coatings, heterogeneous catalysis, supramolecular chemistry, molecular electronics, molecular modeling, lithography for the production of integrated circuits, semiconductor research and quantum dots, quantum computing, MEMS (microelectromechanical systems), liquid crystals, small LEDs, solar cells, hydrogen storage systems, biochemical sensors, targeted drug delivery, molecular biotechnology, genetic engineering, neurophysiology, tissue engineering, and so on".[2]

16.2 Nanotheory as a Cluster of the Different Natural and Engineering Theories

So "Nanotechnology comprises the emerging applications of Nanosciences."[3] Molecular electronics, early recognition of the carcinoma on the molecular level, and paint coating that changes color in response to change in temperature or chemical environment from the experts' assessment are the more long-term prospects.[4] However, the hydrogen accumulation in the nanostructures is estimated as existing

---

[2] Schummer J. Cultural diversity in nanotechnology ethics. In: Interdisciplinary science review, 2006, Vol. 31, No. 3, p. 219.

[3] Schmid G. et al. Nanotechnology. Assessment and Perspectives. Berlin, Heidelberg: Springer-Verlag, 2006, p. 11. Some applications of nanotechnology can be more or less defined. "Nanotubes, depending on their structure, can be metals or semiconductors. They are also extremely strong materials and have good thermal conductivity. The above characteristics have generated strong interest in their possible use in nano-electronic and nano-mechanical devices. For example, they can be used as nano-wires or as active components in electronic devices such as the field-effect transistor shown in this site" (http://www.research.ibm.com/nanoscience/nanotubes.html). "The unusual properties of carbon nanotubes make possible many applications from battery electrodes, to electronic devices, to reinforcing fibers, which make stronger composites. ...we describe some of the potential aims that researchers are now working on. However, for the application potential be realized, methods for large-scale production of single-walled carbon nanotubes will have to be developed. The present synthesis methods provide only small yields, and make the cost of the tubes about $1500 per gram. On the other hand, large-scale production methods based on chemical depositon have been developed for multiwalled tubes, which are presently available for $60 per pound, and as demand increases, this price is expected to drop significantly. The methods used to scale up the multiwalled tubes should provide the basis for scaling up synthesis of single-walled nanotubes. Because of the enormous application potential, it might be reasonable to hope that large-scale synthesis methods will be developed, resulting in a decrease in the cost to the order of $10 per pound" (Ch.P. Pool, Jr., F.J. Owens. Introduction to Nanotechnology. Hoboken, New Jersey: John Wiley & Sons, 2003, p. 125).

[4] "...it is important to recognize that the use of nanostructuring or nanostructures to generate, fabricate or assemble high surface area materials is at an embryonic stage. The effect of the nanostructure and our ability to measure it will be increasingly important for future progress and development of materials for the marketplace" (D.M. Cox. High Surface Area Materials. In: Nanostructure Science and Technology. A Worldwide Study. R&D Status and Trends in Nanoparticles, Nanostructured Materials, and Nanodevices. Final Report. Ed. by R.W. Siegel, E. Hu, M.C. Roco. Prepared under the guidance of the Interagency Working Group on NanoScience, Engineering and Technology (IWGN), National Science and Technology Council (NSTC). WTEC, Loyola College in Maryland, September 1999, p. 61 - http://itri.loyola.edu/nano/final/).

in the stage of the technological realization.⁵ "Current applications of nano-scale materials include very thin coatings used, for example, in electronics and active surfaces (for example, self-cleaning windows). Nano-scale electronic devices currently being developed are sensors to detect chemicals in the environment, to check the edibility of foodstuffs, or to monitor the state of mechanical stresses within buildings. Much interest is also focused on quantum dots, and semiconductor nanoparticles that can be 'tuned' to emit or absorb particular light colors for use in solar energy cells or fluorescent biological labels. Applications of nanoscience and nanotechnologies are also leading to the production of materials and devices such as scaffolds for cell and tissue engineering, and sensors that can be used for monitoring aspects of human health. Many of the applications may not be realized for ten years or more (owing partly to the rigorous testing and validation regimes that will be required). In the much longer term, the development of nanoelectronic systems that can detect and process information could lead to the development of an artificial retina or cochlea. So far, the relatively small number of applications of nanotechnologies that have made it through to industrial applications represent evolutionary rather than revolutionary advances. Current applications are mainly in the areas of determining the properties of materials, the production of chemicals, precision manufacturing, and computing. In mobile phones, for instance, materials involving nanotechnologies are being developed for use in advanced batteries, electronic packaging, and in displays. The total weight of these materials will constitute a very small fraction of the whole product but will be responsible for most of the functions that the devices offer. In the longer term, many more areas may be influenced by nanotechnologies, but there will be significant challenges in scaling up production from the research laboratory to mass manufacturing."⁶ The divided development of physics (electrical engineering – electronics – microelectronics – material design – quant effects), biology (cell biology – molecular biology – functional molecule design), and chemistry (complex chemistry – supramolecular chemistry) in perspective must be integrated in the nano level.⁷

Exactly such a cluster of different theories is nanotechnology as is the most typical representative of the modern technoscience. What integrates all these heterogeneous theories from a large number of different disciplines including physics, chemistry, biology, medicine and engineering sciences⁸? This is only the orientation

---

⁵ H. Paschen, Chr. Coenen, T. Fleischer u.a. Nanotechnologie. Forschung, Entwicklung, Anwendung. Berlin, Heidelberg, New York: Springer, 2004.

⁶ Nanoscience and nanotechnologies The Royal Society & The Royal Academy of Engineering, 2004, p. viii-ix.

⁷ H. Paschen, Chr. Coenen, T. Fleischer u.a. Nanotechnologie. Forschung, Entwicklung, Anwendung. Berlin, Heidelberg, New York: Springer, 2004.

⁸ "As a simple example we can take a Biosensor which allows the detection of DNA sequences by turning the surface Plasmon resonance of nanosized gold particles in a suspension. It can be easily seen that in such a problem *quantum physics, chemistry, biology and finally microtechnology* are involved" (Schmid G. et al. Nanotechnology. Assessment and Perspectives. Berlin, Heidelberg: Springer-Verlag, 2006, p. 440). "Although we were able to develop nonviral gene transfer systems that were efficient enough to gain commercial success *in vitro*, the use of this material *in vivo*

on the general or may be even "universal" world view – "nanoontology." "The most common of these define nanotechnology as the *investigation and manipulation* of material objects in the 1–100 nm range, in order to explore novel properties and to develop new devices and functionalities that essentially depend on that 1–100 nm range. Whether intentionally or not, this definition covers all classical natural sciences and engineering disciplines that investigate and manipulate material objects, including chemistry, materials science, solid state physics, pharmacology, molecular biology, and chemical, mechanical and electrical engineering".[9] Therefore, nanoobjects are identified only with ultimate general ontological properties – with appropriate dimensions irrespective of their nature.

Nanotechnology is not only a new type of scientific-engineering discipline, it also evolves in a "nonclassical" way.

There are two basic methods for the development of "classical" technical sciences: first, from the new applied research directions of any natural-scientific theory; and second, they may "bud out" from a corresponding, basic technical theory within the framework of a "family" of homogeneous scientific-technological disciplines that has recently emerged and that is oriented toward use in the engineering practice and design not only of natural sciences, but also for social sciences and the humanities. The range of design tasks has also been enlarged and now includes the problems of social and economic, engineering, and psychological, systems and other aspects. Finally, there appeared such scientific-technical disciplines as are the result of complicated interdisciplinary processes taking place in the technical scientific areas. Such scientific-technological disciplines may be referred to as the modern complex of ("nonclassical") scientific-technological disciplines. Among them are systems engineering, ergonomics, systems design, informatics, operations research. The present complex scientific-technological disciplines represent a reality of contemporary science. However, they do not fit into the traditional forms of organization or methodological standards. It is precisely the sphere of these investigations and disciplines where an 'internal' interaction of social, natural, and technical sciences is being realized today. These sciences also summarize research and development (R&D) orientations and form a single R&D establishment (for complex research and systems design) in the process of accomplishing complex scientific and technical tasks and solving complex scientific and technical problems. The corresponding investigations, for example, in the field of artificial intelligence, require a special management support and search for new forms of scientific organization (for

---

did not pan out because of the lack of efficiency and untoward effects found in some biological systems. To improve this work and to expand the application of synthetic materials to other applications, *a multidisciplinary approach involving chemists, engineers, and biologists is necessary*" (MANAGING NANO-BIO-INFOCOGNO INNOVATIONS: CONVERGING TECHNOLOGIES IN SOCIETY. Ed. By W.S. Bainbridge and M.C. Roco. National Science and Technology Council's Subcommittee on Nanoscale Science, Engineering, and Technology, 2005. Dordrecht, The Netherlands: Springer, 2006, p. 128).

[9] Schummer J. Cultural diversity in nanotechnology ethics. In: Interdisciplinary science review, 2006, Vol. 31, No. 3, p. 218–219.

instance, in temporary scientific teams and problem groups). To this type of the modern scientific-technological discipline belong nanoscience & nanotechnology.

I. Hacking and R. Giere combine structural view (see Fig. 16.1) with technological approach to the understanding of the scientific theory,[10] which more corresponded to the nano scientific research. Ronald Giere understands theory as a population (family) of models or still better "a family of families of models," which can be related to the reality only indirectly. "A real system is *identified* as being similar to one of the models."[12] In the relation of the theoretical models with the real systems, technology plays a decisive role.

Like Hacking, "Giere's constructive realism sees a proof of reality in the successfully managed technologies in handling entities (e.g. electrons) that earlier had

---

Structure core „K"

$K = < Mp, Mpp, M, C >$

*M*pp is set of ***partial potential Models***

*M*p is set of ***potential Models***

M is set of ***Models*** and

C is set of ***Constraints***

I is set of ***intended applications of theory***, which conditions
I *M*pp must be met.

Basic element of theory „T" is $T = < K, I >$

Hierarchical structure of such elements is the **Network „N"** of the different elements of this theory.

Fig. 16.1 Structural concept of the scientific theory[11]

---

[10] Hacking, I. Representing and Intervening. Cambridge - New York: Cambridge University Press 1983; Giere, R.N. Explaining Science: The Cognitive Approach. Chicago - London: Chicago University Press, 1988.

[11] Sneed J. The Logical Structure of Mathematical Physics. Dordrecht: Reidel, 1971; Stegmüller W. The Structure and Dynamics of Theories. N.Y.-Heidelberg-Berlin: Springer Verlag, 1976.

[12] Giere, R.N. Explaining Science: The Cognitive Approach. Chicago - London: Chicago University Press, 1988, p. 86.

the status of a theoretical entity, if they are applied to cover and characterize new models or other theoretical entities. (Nowadays, we routinely use electron rays in accelerators or in electronic microscopes successfully to accomplish other scientific tasks. We understand in this technological sense the theoretically postulated electrons, which were earlier mere theoretical entities, now as scientific-technological *real entities.*) As electrons and protons are manipulated and applied in big technology measurement instruments and appliances to probe and prove the structure of other elementary particles like gluons, quarks etc., these electrons and protons are now indeed "real."[13] Hence, "some of what we learn today becomes embodied in the research tools of tomorrow."[14]

For example such theoretical entities as quantum dots have found applications in fluorescent biological labels to trace a biological molecule.[15] This application of the quantum dots as marks and contrast agents in the other experiments is a technological verification and an indirect demonstration of the reality of the quantum dots. "Nanotechnologies already afford the possibility of intracellular imaging through attachment of quantum dots or synthetic chromophores to selected molecules, for example proteins, or by the incorporation of naturally occurring fluorescent proteins which, with optical techniques such as confocal microscopy and correlation imaging, allow intracellular biochemical processes to be investigated directly" (Fig. 16.2).[16]

The object of the nanoscience exists first of all only as a computer model that simulates in the definite form the operation of the oncoming system that is to say designer's plan. Scientific investigation is always connected with the computer

---

[13] H. Lenk. Grasping Reality. An Interpretation-Realistic Epistemology. N.J., L., Singapore, Hong Kong, 2003, p. 84–85.

[14] Giere, R.N. Explaining Science: The Cognitive Approach. Chicago - London: Chicago University Press, 1988, p. 140.

[15] "Nanocrystals, also called quantum dots (QD), are artificial nanostructures that can possess many varied properties, depending on their material and shape. For instance, due to their particular electronic properties they can be used as active materials in single-electron transistors. Because certain biological molecules are capable of molecular recognition and self-assembly, nanocrystals could also become an important building block for self-assembled functional nanodevices. The atom-like energy states of QDs furthermore contribute to special optical properties, such as a particle-size dependent wavelength of fluorescence; an effect which is used in fabricating optical probes for biological and medical imaging. So far, the use in bioanalytics and biolabeling has found the widest range of applications for colloidal QDs. Though the first generation of quantum dots already pointed out their potential, it took a lot of effort to improve basic properties, in particular colloidal stability in salt-containing solution. Initially, quantum dots have been used in very artificial environments, and these particles would have simply precipitated in 'real' samples, such as blood. These problems have been solved and QDs are ready for their first real applications" (M. Berger. Quantum dots are ready for real world applications. March 21, 2007. Nanowerk LLC - http://www.nanowerk.com/spotlight/spotid = 1650.php).

[16] Nanoscience and nanotechnologies, The Royal Society & The Royal Academy of Engineering, 2004, p. 23.

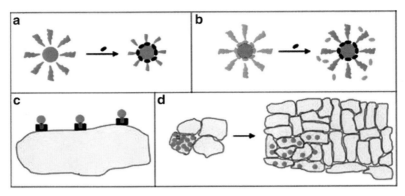

Fig. 16.2 (a) Analyte detection by quenching of the quantum dot fluorescence (red) upon binding of the analyte (black) to the quantum dot surface. (b) By binding an appropriate organic fluorophore (green) as acceptor to the surface of the donor quantum dot fluorescence resonance energy transfer (FRET) occurs. FRET is stopped upon displacement of the acceptor dye from the quantum dots surface by the analyte. (c) Specific cellular receptors (black) can be labeled with quantum dots that have been modified with appropriate ligand molecules. (d) If a cell (grey) within a cell colony is labeled with quantum dots this cells passes the quantum dots to all its daughter cells and the fate of this cell can be observed[17]

simulation and all, what we see in the display, is already determined from some theory and its mathematical representations that are defined in the software of the simulation modeling.

Nanoonthology or nano scientific world view has a function of the methodological orientation for the choice of the theoretical means and methods to a solution to the scientific and engineering problems. This allows to change from one explanation and scientific world view to another without any problems. For example, an electron in one place is considered as a spherical or a point electrical charge, could be rolled spherically symmetrically over the nucleus or as "the free electrons travel through an external circuit wire to the cathode" or in "various electron trajectories". In another place of this book is written: "one can view the electron charge between the two atoms of a bond as the glue that holds the atoms together". In third place we can read that "the electrons in a nanotube are not strongly localized, but rather are spatially extended over a large distance along the tube". And in fourth place of the same book electrons as in the quantum theory can be viewed as waves: "If the electron wavelength is not a multiple of the circumference of the tube, it will destructively interfere with itself, and therefore only electron wavelengths that are integer multiples of the circumference of the tubes are allowed".[18]

---

[17] From: M. Berger. Quantum dots are ready for real world applications. March 21, 2007. Nanowerk LLC - http://www.nanowerk.com/spotlight/spotid=1650.php

[18] Ch.P. Pool, Jr., F.J. Owens. Introduction to Nanotechnology. Hoboken, New Jersey: John Wiley & Sons, 2003, p. 98, 120–121, 128, 243.

In the so-called "teleological" definition, nanotechnology is defined "in terms of future goals. To be specific, one needs to provide more than just generic values, such as health, wealth, security, and so on, and more than just relative attributes like smaller, faster, harder, and cheaper".[19]

16.3 Nano Systems Engineering

Hence, nanotechnology is both a field of scientific knowledge and a sphere of engineering activity, in other words – NanoTechnoScience[20] – similar to Systems Engineering as the analysis and design of large-scale, complex, man/machine, systems but micro- and nanosystems. "Nanoscience deals with functional systems based either on the use of subunits with specific size-dependent properties or on individual or combined functionalized subunits".[21] Nano systems engineering is the aggregate of methods of the modeling and design of the different artifacts (fabrication of nanomaterials, assembling technology for construction of comprehensive micro and nano systems, micro processing technology for realizing micromachines etc.).[22] Nano systems engineering as well as Macro systems engineering includes not only systems design but also complex research. Design orientation has influence on the change of the priorities in the complex research and of the relation to the knowledge, not only to "the knowledge about something", but also to the knowledge as the means of activity: since the beginning, *control and restructuring of matter* at the nanoscale has been a necessary element of nanoscience (Fig. 16.3).[23]

"Nanotechnology is the engineering of functional systems at the molecular scale K.E. Drexler was talking about building machines on the scale of molecules".[24] "Manufactured products are made from atoms, and their properties depend on how those atoms are arranged. This volume summarizes 15 years of research in *molecular manufacturing*, the use of nanoscale mechanical systems to guide the placement of reactive molecules, building complex structures with atom-by-atom control. This

---

[19] Schummer J. Cultural diversity in nanotechnology ethics. In: Interdisciplinary science review, 2006, Vol. 31, No. 3, p. 219.

[20] Discovering the Nanoscale. D. Baird et al. (Eds). Amsterdam: IOS Press, 2005.

[21] Schmid G. et al. Nanotechnology. Assessment and Perspectives. Berlin, Heidelberg: Springer-Verlag, 2006, p. 11.

[22] "Microsystems engineering and nanotechnology are two disciplines of miniaturization in science and engineering, which complement each other. Nanotechnology provides access to so far unused, completely novel effects. Microsystems engineering allows for the development of complete systems solutions due to its highly systemic potentials" (The KIT Nano- and Microscale Research and Technology Center (NanoMikro) – www.fzk.de/fzk/groups/kit/documents/internetdokument/id_059981.pdf).

[23] H. Paschen, Chr. Coenen, T. Fleischer u.a. Nanotechnologie. Forschung, Entwicklung, Anwendung. Berlin, Heidelberg, New York: Springer, 2004, S. 1, 27.

[24] What is Nanotechnology? Center for Responsible Nanotechnology, 2008.

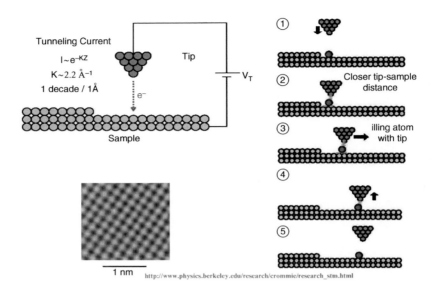

Fig. 16.3 Schematic diagram of molecular manipulation with the Scanning Tunneling Microscope (STM)

degree of control is a natural goal for technology: Microtechnology strives to build smaller devices; material science strives to make more useful solids; chemistry strives to synthesize more complex molecules; manufacturing strives to make better products. Each of these fields requires precise, molecular control of complex structures to reach its natural limit: a goal that has been termed *molecular nanotechnology*. Our ability to model molecular machines – of specific kinds, designed in part for ease of modeling – has far outrun our ability to make them".[25]

Micro-Nano Systems Engineering is quite a new direction in Systems Engineering. Micro/Nano Systems Engineering or systems engineering for micro and nanotechnologies is assembling technology for the construction of comprehensive micro/nano systems, micro processing technology for realizing micromachines, microelectromechanical systems (MEMS), and microsystems. "The microsystems field has expanded to embrace a host of technologies, and microelectronics has now been joined with micromechanics, microfluidics, and microoptics to allow the fabrication of complex, multifunctional integrated microsystems". Micro Systems Engineering is the technologies and capabilities available in this highly interdisciplinary and dynamically growing engineering field: "including design and materials, fabrication and packaging, optical systems, chemical and biological systems,

---

[25] E. Drexler. Nanosystems. John Wiley & Sons, Inc., 1998 - I:\Nano\nano0208\Preface_1.mht

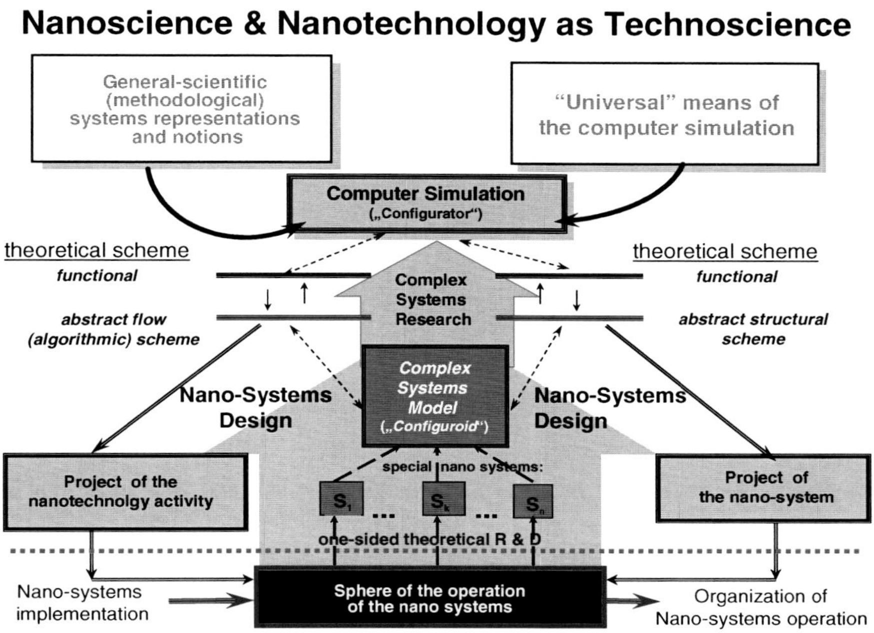

Nanoscience & Nanotechnology as Technoscience

Fig. 16.4 The Structure of NanoTechnoScience

physical sensors, actuation, electronics for MEMS and industrial applications"[26] (Fig. 16.4).

Generalized structural schemes came into being by way of generalizing different structural schemes: automatic control theory, network theory, switching circuit network theory, computer logics, and those used in socio-economic case studies are combined in the so-called structural analysis of complex systems. Such unified abstract structural schemes make it possible to study an object in the most general form. For example, in the course of structural studies of automatic control systems, nothing remains but relations, their number, differential order, sign, and configuration. In nanotechnology, there are quantum circuits (Fig. 16.5) or schematic structure

---

[26] Comprehensive Microsystems, Vol. 1–3. Ed. by Yogesh Gianchandani, Osamu Tabata, Hans Zappe. Hardbound, 2007 - http://www.elsevier.com. See: http://www.me.kyoto-u.ac.jp/micro/english/laboratory/micromachine/micromachine.htm. "Die Mikrosystemtechnik ist eine der Schlüsseltechnologien des 21. Jahrhunderts. Produkte mit mikrosystemtechnischen Komponenten erobern immer mehr Anwendungsbereiche im täglichen Leben und sind in ihren Potentialen hinsichtlich Funktionalität und Wirtschaftlichkeit aus unserem Alltag nicht mehr wegzudenken. Neue Anwendungsfelder werden erschlossen durch Skalierung der Strukturen in den Nanometer-Bereich" (Prof. Dr.-Ing. H. Vogt. Micro and Nano Systems Engineering. - file:///F:/nano/veranst_veranstaltung_473.htm).

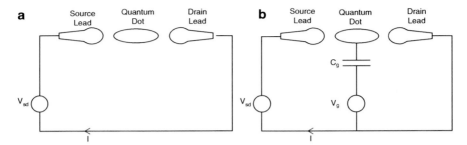

Fig. 16.5 (a) Quantum dot coupled to an external circuit through source and a drain leads. (b) Quantum dot coupled through source and a drain leads to an external containing an applied bias voltage V_{sd}, with an additional capacitor-coupled terminal through which the gate voltage V_g controls the resistance of the electrically active region. Ch.P. Pool, Jr., F.J. Owens. Introduction to Nanotechnology. Hoboken, NJ: Wiley, 2003, p. 245, 246

of the single-electron box[27]. See also as example of such schemes the schematic diagram of the experimental setup of the Single-Atom Transistors – Fig. 11.1(a)[28].

Generalized algorithmic schemes were applied in cybernetics and in the transformation of matter, energy, and information. Actually, they are idealized representations of any system's functioning and are the starting point for computer programming (they are related to the respective functional schemes in the theory of programming). Further manipulation of the model can be done in the simulation languages adequate for the problem. An algorithmic scheme of model (system) functioning is then developed on the basis of the structure given. It is automatically translated into machine code, and in turn, corresponds to a functional (mathematical) scheme. In nanotechnology, such generalized algorithmic schemes can be, for example, the algorithm of lithography (Fig. 16.6) and the algorithm of the transmission electron microscope image processing[29].

The dual orientation of nanotechnology both toward scientific research into natural phenomena and toward production, the embodiment of a conception by artificial means, by purposeful creative work, makes nanotechnology look at any

---

[27] See schematic structure of a single-electron box, consisting of a quantum dot (island), an electron connected to the dot through a tunneling junction, and an electrode coupled to tile dot through an ideal, infinite-resistance, capacitor and equivalent circuit of the single-electron box in: Schmid G. et al. Nanotechnology. Assessment and Perspectives. Berlin, Heidelberg: Springer-Verlag, 2006, p. 149.

[28] See: Ch. Obermair, F. Xie, R. Maul, W. Wenzel, G. Schön, and Th. Schimmel. Single-Atom Transistors: Switching an Electrical Current with Individual Atoms. In this book, p. 115.

[29] See, for example: transmission electron microscope image processing "for a Ni particle on a SiO_2 substrate, showing (a) original bright-field image, (b) fast Fourier transform diffraction-pattern-type image, (c) processed image with aperture filter shown in inset, (d) image after further processing with aperture filter in the inset, (e) final processed image, (f) image of SiO_2 substrate obtained by subtracting out the particle image, and (g) model of nanoparticle constructed from the processed data" (in: Ch.P. Pool, Jr., F.J. Owens. Introduction to Nanotechnology. Hoboken, New Jersey: John Wiley & Sons, 2003, p. 50).

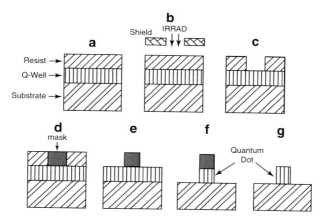

Fig. 16.6 Steps in the formation of a quantum wire or quantum dot by electron-be lithography: (**a**) initial quantum well on a substrate, and covered by a resist; (**b**) radiation with sample shielded by template; (**c**) configuration after dissolving irradiated portion of resist by developer; (**d**) disposition after addition of etching mask; (**e**) arrangement after removal of remainder of resist; (**f**) configuration after etching away the unwanted quantum-well material; (**g**) the final nanostructure on substrate after removal of etching mask

product it develops as a natural–artificial system. On the one hand, nanosystem is a phenomenon that obeys the laws of nature, and on the other, an object that needs to be created artificially (e.g., a nanomachine): "... the focal point of nanotechnology is to produce artificial objects that are more useful for human purposes than natural ones".[30] In turn, the situations artificially embodied in an experiment must themselves be presented and described scientifically as natural processes. Hence, in an experiment of the classical natural science, even the one clearly oriented toward engineering thought, emphasis must be laid mainly on its natural aspect. Traditional engineering, however, emphasizes the artificial aspect, although the engineer himself has a mixed, artificial and natural, attitude. This fact can be explained primarily by the differences between experimental research and engineering practice. The main objective of an experiment is to demonstrate and confirm by artificial means the physical laws derived theoretically, whereas engineering is aimed at developing, based on these laws, the artificial technical means designed to satisfy a specific human need. In nanotechnology, these two positions are so inextricably intertwined that scientific experiment becomes a design of the nanosystems. Thus, if nanotechnology applies scientific knowledge to the creation of nanosystems, it is in the "artificial–natural–artificial" position, but if it develops devices to conduct experiment in order to substantiate this knowledge, it is in the "natural–artificial–natural" position. It means mutual influence of complex research and systems design, which has different functions and plays different roles in NanoTechnoScience.

---

[30] G. Schiemann. Nanotechnology and Nature. On Two Criteria for Understanding Their Relationship. In: HYLE–International Journal for Philosophy of Chemistry, Vol. 11, No.1 (2005), pp. 77–96. - http://www.hyle.org

Index

3D reconstruction, 49

A

acceptor concentration, 83
acoustic driven lab-on-a-chip, 185
acoustic streaming, 171
AFM microdissection, 184
algorithmic scheme, 203
amorphous intergranular film, 79
Andreev reflection, 21
annealing, 79, 119
anodization, 142
artificial blood vessel, 178
atom-by-atom control, 202
Atomic force microscope, 92
atomic force microscopy, 145
atomic-scale contact, 116
atomic-scale junction, 113
atomic-scale reorganization, 120
atomic-scale switch, 117
atomic-scale switches, 113
atomic-scale transistor, 114, 118
automatic control theory, 202
automatized etching system, 139

B

backside illumination, 140
band overlap, 16
Berry phase shift, 15
Bi nanowire, 15
bifurcation, 35, 92
biological reactors, 186
bionic applications, 163
biosensors, 180
bistable switching, 116
blood flow, 177
blood platelet adhesion, 178
bonding energy, 52

borocarbides, 22
boundaries of nanocrystalline grains, 51
boundary diffusion coefficients, 53
branching morphology, 92
bulk and the total conductivity, 83

C

capacity of the tip–sample-arrangement, 150
carbon concentration profile, 48
carbon content, 48
carbon gradient, 48
carbon nanofibers, 103
carbon nanotubes, 103
carbon steels, 41
catalyst particle, 103
Cell–cell interactions, 181
cell-sorting process, 164
cementite decomposition, 45
cementite dissolution, 41, 52
cementite lamellae, 45
cementite morphology, 45
cementite reflections, 46
cementite–pearlite interface, 52
Chambers effect, 16
charge-induced resistance modulation, 131
charged particles, 57
chemical transfer resistances, 141
chemical vapor deposition, 103
chromosomal probes, 182
Cold-drawn pearlitic steel, 50
collection efficiency, 58
collection electrodes, 58
collector stage, 60
commensurate AF order, 25
complex flow pattern, 171
composite structure, 68
compressed nanopowder, 127
computer-controlled feedback, 116
conducting AFM, 156

Conducting oxides, 80
Conductive AFM, 145
Controlled nanomanipulation, 183
copper electrodeposition, 89
Corona discharge, 60
corona discharge suppression, 58
crystal growth, 90
crystal–crystal GB, 79
cubic equation, 32
cyclovoltammogram, 131

D

damping factor, 31
de-alloying, 128
decimation technique, 120
decomposition mechanisms, 50
decreased atomic density, 51
Definitions of Nanotechnology, 194
deformation part, 73
deformation peculiarity, 68
deformed zone, 72
delamination, 73
density of excitons, 32
desegregation, 79
destroyed zone, 71
diagnostic devices, 180
dielectric constant, 30
diffusion channels, 103, 107
diffusive mixing, 186
dislocations, 44, 50
dissolution valence, 142
dissolution/deposition cycles, 117
divalent dissolution, 142
dopant distribution, 150
double layer capacitance, 126
downhill diffusion, 53
drawing speeds, 49
droplet volume, 171
droplet-based assays, 186
dynamic system, 32

E

elastic exciton–exciton interaction, 37
elastic mean free path, 14
elastic–plastic relaxation, 73
electrical double layer, 126
electro-osmosis, 167
electrocapilarity, 129
electrochemical annealing, 119
electrochemical charging, 126
electrochemical cycling process, 114
electrochemical deposition/dissolution cycling, 119

electrochemical double layer, 118
electrochemical fabrication, 114
electrochemical field-effect, 126
electrochemical gating, 129
electrochemical pore formation, 139
Electrocrystallization, 99
electrodeposition, 89
electrodeposition cell, 91
electrokinetic phenomena, 166
electrolytic charging, 127
electromigration, 90, 95
electronic conductivity, 156
electrophoresis, 166, 174
electrostatic force microscopy, 150
electrostatic interaction, 151
electrostatic precipitator, 57, 59
energy-dispersive X-ray spectrometry, 94
enhanced ion-induced dissociation, 109
equilibrium segregation, 79
exchange current density, 119
exchange splitting, 4
exciton density, 31, 33
exciton polarization, 31
exciton self-frequency, 30
exciton–photon interaction, 29
exhaust gas, 57
extended Hückel model, 120
external and integrated actuators, 166
external electrostatic field, 58

F

ferrite matrix, 42
ferromagnetic layer, 7
FFT impedance spectroscopy, 139
field amplitude, 32, 33
field ion microscopy, 43
flexible electronics, 134
flow rate, 162
Flow switching, 164
Flow-through PCR-chips, 173
fluidic environment, 172
flux quantization, 19
forensic applications, 185
fractional collection efficiency, 62
fractional particle number collection efficiency, 64
Fuchs–Sondheimer model, 132
Fuel cells, 145

G

galvanostatic mode, 98
gap suppression, 23
gate electrode, 114

Index

Gauss-like incident pulse, 35
GB conductivity, 82
GB diffusion, 79
GB network, 81
GB segregation, 77
GB segregation layer, 84
GB triple junctions, 80
GB wetting conditions, 81
GB wetting layer, 77
GB wetting transforms, 78
Generalized algorithmic schemes, 202
genetic analysis, 180
glassy phase, 80
globular character, 69
globules, 69
globules flattening, 69
grain boundaries, 44
grain refinement, 85
graphene, 104
graphene–catalyst interface, 108, 109

H

Hückel model, 120
half-widths, 35
handling entities, 198
hard-rigid monocrystalline solids, 68
high shear flow, 179
high-resolution imaging, 184
homogenous laminar flow, 177
hydrodynamic models, 179
hydrodynamic switches, 164
hydrodynamic system, 169
hydrophilic track, 178
hydrophillic regions, 172
hydrostatic pressure, 50
hysteretic dependence, 32

I

impedance spectroscopy, 84
incident amplitude, 33
incident pulse, 36
incommensurate, 27
incommensurate spin density wave, 25
indentation contact zone, 69
indentation depth, 71
indentation-deformed zone, 73
indenter penetration, 71
independent control electrode, 114
Indium tin oxide, 133
inert gas condensation, 127
integrated nanopump, 172
integrated planar pumps, 172
integrated silicon microchamber array, 176
interdigital transducers, 166
interdigitated transducers (IDT), 170
interface transparency parameter, 8
interference conditions, 5
intergranular phase, 77
interlamellar spacing, 48
intermittent contact mode, 152
internal acoustic streaming, 186
internal streaming, 171
internal stresses, 72
interphase boundary, 53
ion-assisted dissociation, 109
ionic conduction, 84
ionic conductivity, 156
ionic–electronic conductors, 145
ionizing stage, 60
isotropic Fermi surface, 23
ITO nanoparticles, 133

J

Jacoby function, 36
junction field-effect transistor, 126, 134

K

Keldysh equation, 30

L

lamellar structure, 41
laser-based microdissection, 182
laser-photoemission spectroscopy, 25
lattice expansion, 51
lattice parameter, 51
leakage current, 142
lift mode, 152
liquid aerosol, 60
liquid ceramics, 86
liquid ceramics technology, 85
liquid-phase sintering, 77
localized charges, 151
longitudinal MR oscillations, 14

M

macropore growth, 139
magnetic coherence length, 4
magnetic field derivative, 16
magnetic flux, 14, 18
magnetic state, 27
magnetic susceptibility, 129
magnetically ordered state, 27

magnetron sputtering, 6
mass collection efficiency, 61
mass density, 169
mean field approximation, 30
mechanical pumps, 165
metal organic deposition, 146
metal–electrolyte interface, 134
metal–superconductor contact, 22
metallic alloys, 44
metallic quantum point contacts, 114
metastable compound, 41
micro flow chamber, 180
micro total analysis systems, 168
Micro-Nano Systems Engineering, 202
microchannels, 163
microdissection, 180
microfabricated structures, 163
Microfluidics, 161
microhardness, 45, 68
Micropumps, 165
microscale fluidic networks, 162
microstructure, 45, 68
microstructure of film surfaces, 70
microsystems, 202
Microwires, 92
modular lab-on-a-chip system, 185
molecular cytogenetic studies, 182
molecular cytogenetics, 182
molecular manufacturing, 202
molecular nanotechnology, 202
Monolayer GB Segregation, 81
Moreover, for a long-wavelength oscillation, 6
morphology, 90
morphology of copper deposits, 96
mount and the holding clamp had, 152, 156
multiband behavior, 26
multigap superconductivity, 27
multilayer segregation, 75
multilayer systems, 97
multiple reentrant behavior, 9

N
n-type Si, 139
nano scientific world view, 200
Nano systems engineering, 201
Nano-scale electronic, 196
nano-wiring, 89
nanochemical reactions, 172
nanocrystals, 47
nanometer grain structure, 85
Nanoonthology, 200
Nanoontology, 193, 194
Nanoporous Gold, 130

nanoporous metal, 128
nanoporous structure, 126
nanoscale lamellar structure, 50
nanoscale particles, 44
Nanostructured conducting oxides, 75, 86
nanostructuring of ferrite, 45
Nanotechnology, 194
NanoTechnoScience, 193
nanotubes, 18
natural–artificial system, 204
natural–artificial–natural position, 205
Navier–Stokes-Equation, 169
Neel temperature, 21, 25
network theory, 202
nickel-borocarbide, 21
nonadsorbing electrolyte, 133
nonaqueous electrolyte, 130
noncontact heating, 174
nonequlibrium phases, 85
nonlinear optical properties, 30
nonlinear solution, 36
nonlinearity, 30
Nonmechanical pumps, 166
nonmechanical pumps, 165
nonstationary behavior, 33
nonstationary transmissions, 37
Not only the conductance observed by closing the, 117
Nyquist plot, 140

O
Off-stoichiometric cementite, 53
off-stoichiometry cementite, 48
one-gap approach, 24
one-dimensional nanomaterial, 41
optical bistability, 29
oscillations of exciton density, 33
overall conductivity, 84

P
partial plastic flow and material, 69
particle charging, 61
particle mass concentration, 64
Particle number concentration, 63
pearlite colonies, 50
pearlitic steel, 42, 48
periodical nanostructured films, 97
Perovskite-type oxides, 84
phase diagrams, 75
phase shift, 18
phase transfer pumps, 167
phase transformations, 75

Index

photoemission spectroscopy, 17
photogenerated holes, 141
photoinduced dissolution, 141
piezoelectric substrate, 171, 175
pile-up zone, 72
plasma-assisted growth, 104, 109
plasma-enhanced, 104
plastic–elastic recovery, 72
polarization, 29, 30, 156
polymerase chain reaction, PCR, 172
pore etching, 139
positional cloning, 182
powder sintering, 86
printable macroelectronics, 136
programmable fluidic microprocessor, 161
programmable microfluidic chip, 176
Proteins under Flow, 179
purification effect, 83

Q

quantum devices, 122
quantum dots, 199
quantum interference effects, 14

R

Rashba SO interaction, 17
Rashba spin-orbit interaction, 19
redistribution of carbon atoms, 50
redistribution of stresses, 73
reentrance, 6
reentrant behavior, 23
reentrant behavior of superconductivity, 9
reentrant superconductivity phenomenon, 10
research and, 197, 198
resistance, 84
resistance measurements, 8
resonant laser radiation, 30
Reynold's number, 169
Rutherford backscattering spectrometry, 7

S

S/F proximity effect, 4
saddle, 32
SAW-driven microfluidic chip, 177
scaling up, 196
scientific-technological disciplines, 197
secondary phase, 79
segregation effect, 90
self-consistent calculation, 132
self-organization, 97
serpentine channel, 175

Sharvin conductances, 8
short electric pulses, 43
silver point contacts, 115
Single-atom transistors, 113
single-gap approach, 23
small fluid volumes, 161
small-amplitude oscillations, 33
solid electrolyte, 80
solid oxide fuel cells, 145
source-drain conductance, 116
spatiotemporal oscillations, 93
Stationary On-chip PCR, 173
stationary single chamber, 175
stationary solutions, 32
stationary values, 32
strain rate, 49
strain-induced cementite dissolution, 42
subthreshold swing, 134
superconducting coherence length, 8
superconducting spin switch, 10
superconductor–ferromagnetic metal (S/F) contacts, 3
supershort pulses, 29
surface acoustic wave, 166
surface acoustic waves, 170, 175
surface carrier density, 18
surface charge carriers, 17
surface charge density, 126
surface diffusion flux, 105
switching circuit network theory, 202
Systems Engineering, 201

T

technoscience, 194
temperature gradient, 175
The mean field penetration, 126
theoretical entities, 199
thermal convection, 175
thermal degradation processes, 145
thin semiconductor films, 29
tight-binding-like conductance calculations, 113
time evolution of the exciton density, 34
time evolution of the system, 33
tissue engineering, 163
transmission, 29
transmission function, 35
transmittance, 37
transmitted pulse, 35
transmitted radiation, 33
tunneling spectroscopy, 26
two-gap approach, 24
two-dimensional patterns, 95

U

Ulitovsky fabrication process, 15
Ulitovsky technique, 15
ultrafine particle collection, 57
Ultrasonic pumping, 165
ultraspeed control, 29
Ultrathin films, 127
ultrathin layer, 92, 95
ultrathin layer electrodeposition, 100
Under the optimal pore growth conditions this is solely, 141
unipolar particle charging, 64
upper critical field, 23

V

Van der Pauw method, 131
variable thickness, 6
velocity field of the flow, 169
virtual liquid test tube, 176

virtual reaction chamber, 175
viscosity, 169

W

water layer, 157
wave-like morphology, 97
weak localizations, 19
weak pulse, 35
wedge samples, 7
Wire Formation, 95

Y

yttria-stabilized zirconia, 147

Z

zero bias conductance, 120